Technical Guide to GMRS

By
Sean D. Liming and John R. Malin

Annabooks®

Copyright

Technical Guide to GMRS

Published in the United States by

Annabooks, LLC.
85557 Veneto Ln
Indio, CA 92203
www.annabooks.com

ISBN-13: 979-8-9854172-5-8

Dedication

Dedicated to all Radio Geeks Everywhere

Table of Contents

Preface

Growing up I had a fascination with electronics. A Radio Shack electronic kit was given to me as a Christmas present, and I dove into it immediately. Building a crystal radio and a little transmitter was fun. Eventually, I turned to computers and programming as I got older. My college degree was in Electrical Engineering with a focus on Computer Engineering. I could have kicked myself for not taking the extra class that delt with radio transmission, but I have been doing well in my career regardless.

My first introduction to General Mobile Radio Service (GMRS) was at a Microsoft Windows Embedded conference in Beijing, China in 2004. A pair of Uniden GMRS-1000 radios were given as a speaker gift. The radios were to be used to communicate with conference technical staff if I ran into any problems during the presentation. They also warned that an FCC license would be needed to use them in the USA. I used the radio once during one of my presentations, brought them home, and completely forgot about them.

It is not all sunshine and surf living in Southern California. Earthquakes and wildfires are a common occurrence. Several years ago, I was sitting in a golf cart on the 5th fairway at Black Gold Golf Course in Yorba Linda when the ground started to shake. My mom was home alone, and I gave her a call on my cell phone to make sure she was okay. She was fine, but what if I couldn't get through?

A few years later, we moved to Rancho Mirage, CA near Palm Springs, which is closer to the famous San Adreas fault. The home owner's association (HOA) created an emergency response group. Part of the response was radio communication. Because of my regular job, I didn't join right away, but I remembered the Uniden radios that were stuck in a drawer for two decades. During COVID, I researched GMRS, got a GMRS license, and purchased a couple of GMRS radios. Connecting to the local GMRS repeater was easy. Through the repeater, I was able to reach someone at the other end of the valley 30 miles away. Eventually, I joined the emergency communications team. All the other members had HAM licenses, but not GMRS. I got my HAM license, and a few other on the team got their GMRS license.

During my GMRS research, it became clear there GMRS was curiosity for HAM operators, but why the curiosity was an open question. GMRS is limited compared to HAM radio frequencies available. The fact that antenna could be replaced and the tools that amateur radio hobbyists could be used, sparked an idea for this book. Of course, that one class and lab in college I didn't take will make it a challenge, but it is all learn by doing anyways.

Sean D. Liming
Rancho Mirage, CA

GMRS: WRKT830
HAM: KN6YZM

Acknowledgments

We would like to thank BuyTwoWayRadios, R & L Electronics, and Retevis for their timely shipping and faster responses to our questions.

Annabooks

Annabooks provides a unique approach to embedded/IoT system services with multiple support levels. Our different offerings include books, articles, training, and project consulting. Current publications and courses have focused on embedded PC architecture and Windows Embedded, which reach a wide audience from Fortune 500 companies to small organizations. We will continue to expand our future services into new technologies and unique topics as they become available.

Books
Starter Guide for POS Device Applications Using .NET
Starter Guide for STM32Cube™ and Eclipse ThreadX®
Starter Guide to Windows 10 IoT Enterprise, 2nd Edition
Java and Eclipse for Computer Science
Open Software Stack for the Intel® Atom™ Processor
Real-Time Development from Theory to Practice
The PC Handbook

Training Courses
Windows® 10/11 IoT Enterprise Training Course

Please contact us for more information on consultation and availability.

Annabooks®
Web: www.annabooks.com

1 Staying Connected

We human beings like to stay connected to one another. Human history has shown our ingenuity to make a connection: smoke signals, lanterns, messenger pigeons, messenger runners, pony express, telegraph, telephone, television, and satellites. Almost everyone has a smart phone and some internet communication. We have become dependent on our technology to keep us connected with one another. What happens when there is no cell phone service or a disaster strikes and communications goes down? Smart phones are not going to be useful. Falling back to basic two-way radio communication becomes important. Security guards, warehouses employees, movie sets, home improvement stores, and concert roadies rely on two-way radios to communicate over the short distances.

With so much attention on artificial intelligence, computer games, and smart phone apps, amateur radio enthusiasts are keeping the old radio technology alive and an option when other technologies are not available or practical. The Federal Communications Commission (FCC) regulates the radio spectrum in the United States, and the FCC has allocated frequencies for public and amateur radio use. This book will focus on the publicly available frequencies for General Mobil Radio Service (GMRS).

1.1 GMRS History

FCC commissioned Class A and Class B Citizens Radio Service in the 1940s. The Class A service would be the predecessor to GMRS. For Class A, the transistor tube transceivers were limited to 60 watts output while the Class B was limited to 5 watts. The Class A service was FM and transmitted on UHF frequences around 462 MHz. There was 50kHz channel spacing and a ±15 kHz transmitter deviation and. The Class A service was licensed by business.

The UHF 450-470MHz frequency range was modified in the 1960s. Transistors were replacing vacuum tubes making radios more affordable. The channel spacing was reduced to 25kHz with a deviation change to ±5Khz. The change effectively doubled the available channels in the frequency range. In the 1970s, the output wattage was reduced to 50 watts. The more companies using the spectrum caused congestion so in 1987, business license were discontinued. The FCC opened up the business radio service channels for business t use. The current spectrum was renamed to General Mobile Radio Services (GMRS) where individuals can pay for a license.

The Family Radio Service (FRS) was created in 1996 as a low power FM that didn't require a license. Radio Shake success proposed the service so families had a short-range radio to communicate with one another. This was before cell phones became popular. FRS operated in the same frequency range as GMRS. In 2017, the FCC solidified that the FRS will operate on the same 22 channels as GMRS, but not the repeater channels. FRS will only have a maximum output power of 2 watts, and the antenna must be fixed to the radio. GMRS radios maximum output power remains at 5 watts for channels 1-7 and 50 watts for channels 15-22. Most important is that the antenna can be changed for a GMRS radio. The GMRS frequencies are limited to 20 kHz bandwidth.

FRS doesn't require a license. GMRS requires a license, and the license is good for you and your immediate family. The license fee and duration have changed over the past decade. As of this writing, the current license fee is $35, and the license is good for 10 years.

Finally, the most recent change to GMRS is that GMRS repeaters cannot be linked to one another over the internet.

1.2   GMRS Uses and the Gateway to Amateur Radio

GMRS popularity has grown over the years. GMRS repeaters extends the communication range for several miles depending on the location of the repeater. GMRS is used in a variety of activities:

- Family farms
- Hunting
- Car caravans and road trips
- Emergencies
- Local or rural communities

- Hiking
- Biking
- Camping
- Family outings
- Shopping for a Christmas tree at a Christmas tree farm

With the ability to change the antenna and reach repeaters, GMRS is similar to Amateur Radio (HAM). HAM operators use two-way radios to communicate over UHF and VHF frequences either point-to-point or through a repeater. The same amateur radio test equipment to check antenna compatibility and radio performance can be used for GMRS radios.

GMRS is limited to 22 channels and 8 repeater channels. HAM radios licensees have several frequency bands available to them from 1.8 MHz to 1.3 GHz. To access all these frequencies, you have to take a test(s) for a HAM license. Since GMRS doesn't require a test, GMRS provides a cheap way to learn and use radio communications and tools, and an avenue to pursue a HAM license to access a wider range of communication frequencies. GMRS is a benefit to HAM radio operators since not every family member wants to get HAM license. In an emergency, GMRS radios make much more sense since it is one call sign for everyone in the immediate family. Family members can reach each other or other local GMRS operators, while HAM operators can reach to the outside world.

1.3   About the Book

The gateway to amateur radio idea is the driving force behind this book. This chapter will end with getting a GMRS license. Chapter 2 will cover the basics for radio and the different radio services. Chapter 3 covers the GMRS specific frequencies, privacy tones, and terminology. Chapter 4 looks at the different types of GMRS radios, manufacturers, and a deep dive into a few of these radios. Chapter 5 covers working with repeaters. Chapter 6 will dig into the different test tools that can also be used with GMRS and HAM radios.

1.4   Getting the GMRS Licensing

You can buy a GMRS radio and listen all you want, but if you want to talk, you will need a license. Even if you have a HAM Extra Class license, you still need to a GMRS license to transmit on GMRS frequencies. Getting a GMRS license is a two-step process. First is to get an FRN number, and once you get the FRN number, you can than signup for the GMRS license and pay your fee.

1. Open a web browser and go to the FCC Commission Registration System (CORES) site: https://apps.fcc.gov/cores/userLogin.do?
2. The first step is to register with the site so click on the Register button

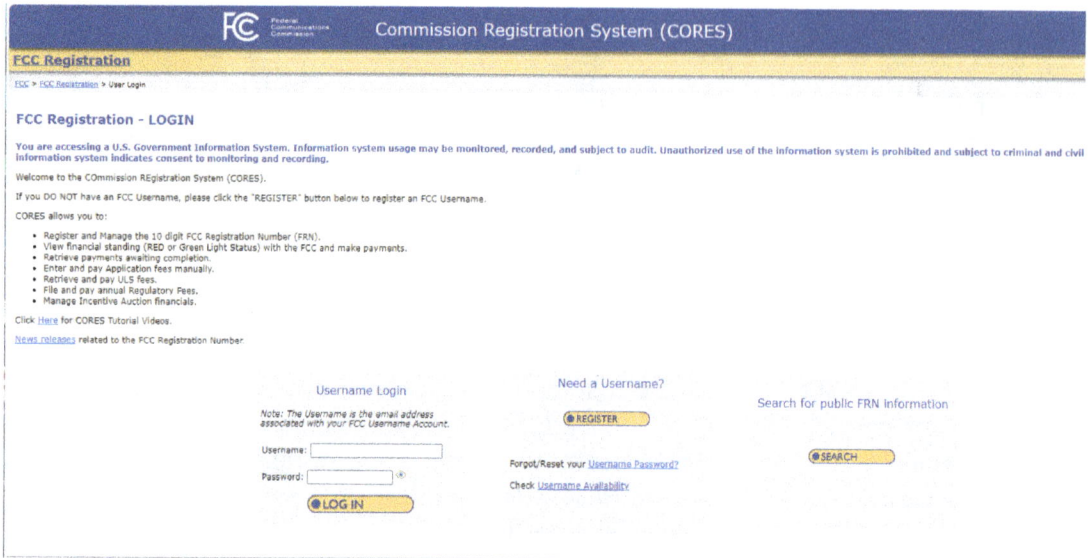

3. Fill out the form and click "Create Account".

Note: The password will be used with the FRN number to get the GMRS license later.

4. In the next screen select "Register New FRN."
5. Select individual, confirm your address is in the USA, and click Continue.
6. There will be one more form to fill out, and click Submit. You will get the FRN number. This number will be important later if you want to get a HAM license. You can also go back to the system and do a search for your FRN number if you lose it.
7. The second step is to go the Universal Licensing System (ULS) website: https://wireless2.fcc.gov/UlsEntry/licManager/login.jsp.

8. Enter your FRN and password created earlier when you registered a new account with the FCC. Click Submit.
9. In the next page, click on "Apply for A New License" from the menu in the top left.
10. In the Select Service drop down select "ZA –General Mobile Radio" (GMRS)" and click Continue.
11. You will have to fille out a Felony Questions form. Click Continue.
12. A summary will appear. If anything needs to be edited, you can do so, or just click on "Continue to Certify".
13. You will then submit the application and will have to pay the fee. Click Continue for Payment Options and select your payment method.

It may take 24 hours, but an e-mail will tell you the license is ready.

14. You will have to log back into the ULS page to download the license.
15. Once you log in, you will see the licenses assigned to you including your call sign. Click on the call sign.
16. You will get to the license at a glance page. There will be your contact information, the license grant date, and the license expiration date. In the banner there is a link to download the license.

Note: Be sure to keep your contact information up to date in the FCC database. You can renew the license within 90 days of expiration.

1.5 Book Page and Feedback

No one likes seeing mistakes in their radio manual, and the same goes for a technical book. Unfortunately, errors sneak through. Books are static compared to webpages. Any errata will be posted to the book page at www.annabooks.com. Feedback is welcome if you think something is missing or a topic needs to be covered more in-depth.

1.6 Summary: Keep Connected

Technology continues to evolve, but sometimes going back to the basics can be a life saver when modern technology fails. Basic radio communication hasn't changed much, but the types of radios and radio options have improved. There are different radio services available, but GMRS provides higher power radio solution to connect with others within a few-miles radius. GMRS radios can reach a longer distance when connected through a GMRS repeater. Even if you just buy the radio and keep it in the car, you will have a device that can keep you connected when cell towers fail.

2 Radio Basics and Radio Services

Before we dive into GMRS and play with the radios, understanding the basics of how radio technology works will help clarify the radio services and their frequence allocations today. In this chapter, we will cover a basic understanding of electromagnetism, discuss the radio frequency spectrum, and the different radio services that are available.

2.1 *Electricity and Magnetism Relationship*

Radio is basically the application of electromagnetism. Through the late 1700's into the 1800s the understanding of electromagnetics was improving. Charles-Augustin de Coulomb, André-Marie Ampère, Michael Faraday, Emil Lenz, and Franz Ernst Neumann laid the ground work within 50 years. The work of James Clerk Maxwell brought everything together in his famous Maxwell Equations. The application of these equations was the final step. Nicola Tesla made strides with radio technology, but it was Guglielmo Marconi who is credited with the first radio transmission. Today, the radio technology has advanced to create radar, cell phones, WiFi, Bluetooth, and NFC.

Rather than covering calculus, differential equations, physics, chemistry, and all the course for an electrical engineering degree, the discussion here will be kept as simple as possible.

2.1.1 Electromagnetic Field

What these early pioneers discovered was that a charge particle develops and electric field. The more the charge the greater the electric field. When the charged particle is at rest, the particle has a static charge. Think of rubbing your feet on a rug and then touching a mettle to get shock. When the particle is moving, a new force appears, magnetism. The electric and magnetic fields act orthogonal to each other, and they radiate out in all direction from the moving particle.

Electric Field

Y

Wavelength λ

Magnetic Field

Z

X

Direction

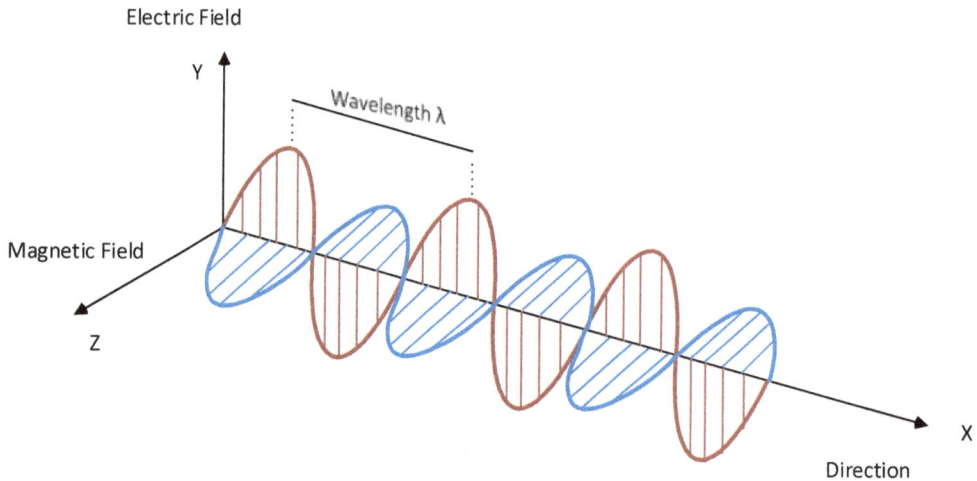

A simple elementary example is to wrap a nail with a copper wire. Connect both ends of the wire to a battery and the nail becomes a magnet. Add a switch and put the nail near an AM radio. Tune the radio to an empty static frequency, and then toggle the switch. A thumping sound comes out the radio. The electromagnetic wave propagates across the air and excites the AM radio antenna, which amplifies the sound. The nail is a simple AM radio transmission demonstration that only broadcasts a few inches.

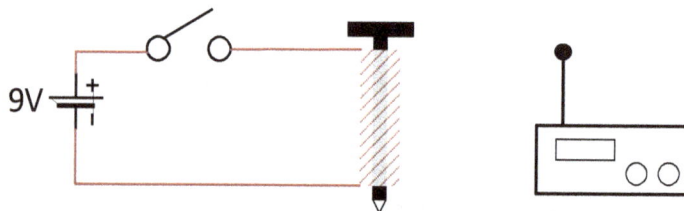

9V

2.1.2 Carrier Frequency and Bandwidth

The nail-battery-switch circuit is a simple demonstration, but a real broadcast station is much more sophisticated and powerful to send a signal over a longer distance. To send voice and data, the information is modulated on to a carrier frequency for transmission. There are three types of modulation: amplitude modulation (AM), frequency modulation (FM), and pulse modulation. GMRS uses FM to modulate the voice when talking into the radio. When you are tuning to a radio station with an AM/FM radio, you are filtering out all other frequencies using a band pass filter to center on the frequency you want to listen to. The filter attenuates all other frequency for a certain bandwidth available to the broad cast frequency. The bandwidth of all the radio stations has a limit set by the local government agencies. In the US, AM broadcasting radio station have a limited of 20kHz bandwidth. FM broadcasting radio stations have a 0.2 MHz bandwidth which creates 100 channels between 88.1 MHz to 107.9 MHz. GMRS has 20kHz bandwidth to separate the channels.

2.1.3 Power and Line of Site

In the nail-battery example, the transmission could only be made in a few inches. The more energy or power (Watts) added, the further out the electromatic wave can travel and the further two radios can be away from each other and still communicate.

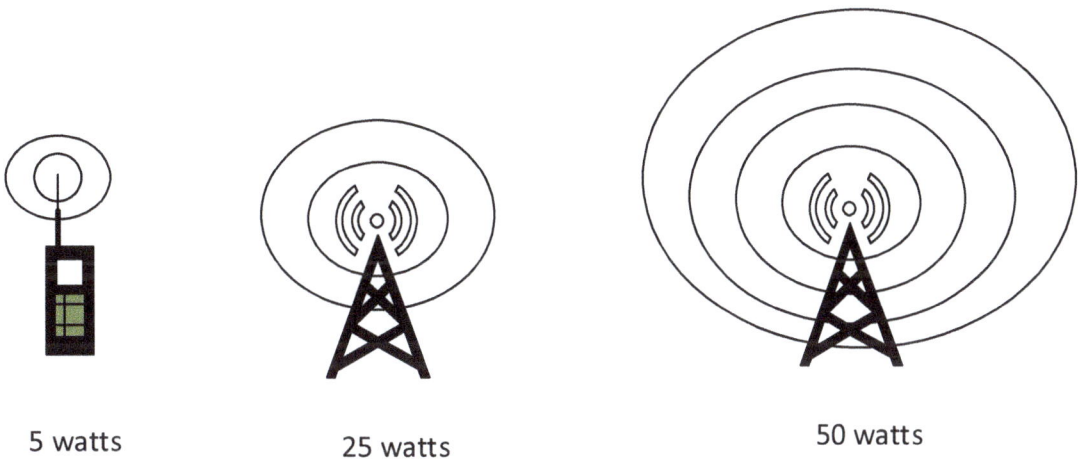

5 watts   25 watts   50 watts

GMRS hand held radios are limited to 5 Watts since the antenna is close to the human being. Getting too close to a high-power antenna is not good for one's health. The power for GMRS mobile radios and repeaters is up to a max limit of 50 Watts. Radio waves can be absorbed by buildings and terrain. The radios should have a good line of site of each other in order to communicate. Repeaters on installed in a high place like build or a mountain, can allow for two radios on either side communicate with each other.

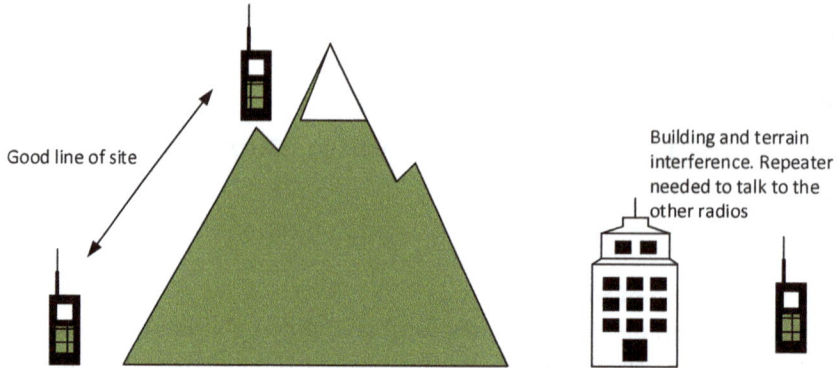

Good line of site

Building and terrain interference. Repeater needed to talk to the other radios

2.2 Frequency Spectrum and Wavelength

In the early days of radio, one could broadcast on any frequency. To monetize the frequency spectrum and prevent stations collide into each other, each country created their own agency to allocate and manage the frequency spectrum. The Federal Communications Commission (FCC) was formed in the United States in 1934. Table of Frequency Allocation chart PDF is available on the FCC website: https://www.fcc.gov/engineering-technology/policy-and-rules-division/radio-spectrum-allocation/general/table-frequency. A graphical chart can be found on the NTIA website: https://www.ntia.gov/page/united-states-frequency-allocation-chart.

Radio frequencies are divided into ranges:

Extremely low frequency	ELF	3Hz -30 Hz
Super low frequency	SLF	30Hz -300Hz
Ultra-low frequency	ULF	300 Hz-3 kHz
Very low frequency	VLF	3 kHz-30 kHz
Low frequency	LF	30 kHz -300 kHz
Medium frequency	MF	300 kHz-3 MHz
High frequency	HF	3 MHz – 30 MHz
Very high frequency	VHF	30 MHz-300 MHz
Ultra-high frequency	UHF	300 MHz – 3 GHz
Super high frequency	SHF	3 GHz – 30 GHz
Extremely high frequency	EHF	30 GHz – 300 GHz
Terahertz high frequency	THF	300 GHz – 3THz

AM radio broadcasts fall into the MF range. FM broadcasts fall into the VHF range. GMRS falls into the UHF range.

The FCC encourages amateur radio by providing access to specific bands across the spectrum. Speaking the long frequency numbers for ranges can be cumbersome. Amateur radio (HAM) operators talk about the bands in terms of wavelength. Frequency and wavelength are converted into one another using the following formula.

$$\text{Wavelength } (\lambda) = c/\text{Frequency } (f)$$

λ = wavelength in meters
c = speed of light in a vacuum (~299,792,458 m/s for calculation some round to 300,000,000m/s)
f – frequency in Hertz

For example, the 1.8 MHz to 2 MHz bad is known as the 160 Meters band.

Wavelength = 300,000,000 /1,800,000 = 166.7 meters.

There is some rounding going on here. In terms of wavelength, GMRS is 65 cm band.

Wavelength = 300,000,000 /462,000,000 = 0.649 meters.

2.3 Different Radio Services

From the radio allocation chart, the spectrum has been divided up for commercial, private, and public use. There are several radio services available for public use.

2.3.1 Citizen Band (CB)

CB radio was made popular in the 1970. Movies like *Smoky and the Bandit*, and TV shows like *Dukes of Hazard, BJ and the Bear*, and *The Misadventures Sheriff Lobo* popularized CBs in the 1970 and into the 1980s.CB radio goes back to the 1940s like GMRS. Since UHF radios were expensive at the time, a Class D of the Citizens' Radio Service was formed for a cheaper HF radio. CB frequency range is split into 40 channels between 26.9650 MHz and 27.4050 MHz. The max power output is limited to 4 Watts. For those old enough to remember, the Archer Space Patrol walkie talkies transmitted channel 14 (27.1250 MHz).

2.3.2   Radio Controlled (RC) Vehicles

Not part of the Citizens Radio Service but provided as a reference, RC toys (cars, boats, and aircraft) are allocated frequencies for communications.

Frequency	Channels
27 MHz	26.995 MHz - Ch 1 (Brown)
	27.045 MHz - Ch 2 (Red)
	27.095 MHz - Ch 3 (Orange)
	27.145 MHz - Ch 4 (Yellow)
	27.195 MHz - Ch 5 (Green)
	27.255 MHz - Ch 6 (Blue)
49 MHz	49.830 MHz - Ch 1
	49.845 MHz - Ch 2
	49.860 MHz - Ch 3
	49.875 MHz - Ch 4
	49.890 MHz - Ch 5
50 MHz (Requires a HAM license)	50.800 MHz - Ch RC00
	50.820 MHz - Ch RC01
	50.840 MHz - Ch RC02
	50.860 MHz - Ch RC03
	50.880 MHz - Ch RC04
	50.900 MHz - Ch RC05
	50.920 MHz - Ch RC06
	50.940 MHz - Ch RC07
	50.960 MHz - Ch RC08
	50.980 MHz - Ch RC09
72 MHz (RC aircraft)	50 Channels – 72.010 MHz – 72.990 MHz
75 MHz (surface RC only. Not for aircraft)	30 Channels – 75.410 MHz – 75.990 MHz

2.3.3   Multi-Use Radio Service (MURS)

MURS was created in 2000. MURS is intended for both personal and business use. MURS comprises of 5 channels on 5 frequencies. The first three channels are limited to narrow bandwidth of 11.25 kHz. The last two channels allow for wide bandwidth of 20 kHz.

Channel	Frequency	Maximum bandwidth
1	151.82 MHz	11.25 kHz
2	151.88 MHz	11.25 kHz
3	151.94 MHz	11.25 kHz
4	154.57 MHz	20.00 kHz
5	154.60 MHz	20.00 kHz

2.3.4   Amateur Radio HAM

HAM radio has several frequency band allocations available for voice, data, and morse code (CW) transmission. A license is required to broadcast on any of the frequencies allocated for amateur radio. You have to take a test to get a license. There are currently 3 license levels: Technician, General, and Extra. Each level opens access to more of the amateur allocated spectrum. To get to the Extra level, you have to take 3 tests. There are books that provide the questions and answers to the test so it is not difficult to get a license.

The frequency bands span from MF to UHF. Besides, the license requirements. certain bands have max power or transmission type limitations.

Band	Frequency	Range
160 Meters	1.8 MHz – 2.0 MHz	LF
80 Meters	3.5 MHz – 4.0 MHz	HF
60 Meters	5.332, 5.348, 5.3585, 5.373, 5.405 MHz	
40 Meters	7 MHz – 7.3 MHz	
30 Meters	10.1 MHz – 10.150 MHz	
20 Meters	14 MHz – 14.350 MHz	
17 Meters	18.068 MHz – 18.168 MHz	
15 Meters	21.0 MHz – 21.450 MHz	
12 Meters	24.890 MHz – 24.990 MHz	
10 Meters	28.0 MHz – 29.7 MHz	
6 Meters	50.0 MHz – 54 MHz	VHF
2 Meters	144.4 MHz – 148.0 MHz	
1.25 meters	222.0 MHz – 225.0 MHz	
70 cm	420.0 MHz – 450.0 MHz	UHF
33 cm	902.0 MHz – 928.0 MHz	
23 cm	1.24 GHz – 1.3 GHz	

There are different radios to handle the different bands. For MF and HF, the radios can bounce the signal off the atmosphere for communications across the globe. VHF and UHF radios communicate via line of site as those frequencies break through the atmosphere. Repeaters can be used to extend the distance for VHF and UHF radios.

2.3.5 FRS/GMRS

The last radio services is the topic of this book. FRS and GMRS use the same frequence. As discussed in the last chapter FRS is limited in power, radio specifications, and cannot connect to repeaters. FRS doesn't require a license. GMRS requires a license, but there is no test like HAM radio. FRS/GMRS frequency range is split into 22 channels. The frequency range are 462.5625 MHz to 462.7125 MHz, 467.5626 MHz -467.7125 MHz, and 462.5500 MHz – 462.7250 MHz. There are 8 repeater frequency 467.5500 MHz to 467.7250 MHz. The next chapter covers the frequencies and channels in more detail.

Compared to HAM radio, GMRS is limited to specific frequencies like CB and MURS. Like HAM VHF and UHF radio transmission, GMRS allows for repeaters so there are some similarities. CB and MURS don't allow for repeaters. HAM radio operators get access to more spectrum depending on their license. What confuses some HAM operators is that they have access to 420.0 MHz to 450.0 MHz, which is near the GMRS frequencies, but they still require a GMRS license to transmit on GMRS.

	HAM	GMRS
Frequencies	Several bands available across HF, VHF, and UHF	Limited to 22 channels and 8 repeater channels in UHF
Licensed required	Yes – A per person license, there are three license levels: Technician, General, Extra	Yes – one license and allows anyone in the immediate family to use the call sign
Test Required	Yes - there is a test for each license level	None
Repeater access	Yes	Yes
Radio types	Base station, hand held, mobile, repeater	hand held, mobile, repeater
Antenna	Different antennas can be attached to the radio	Different antennas can be attached to the radio

To add to the confusion, there are programmable UHF radios that can be configured for 400 MHz to 470 MHz. There are frequencies between HAM and GMRS that can be used for on-premise land communication use.

2.4 Summary:

The discover of the electromagnetic wave opened up a new way to communicate, but also forces countries to regulate the frequency spectrum. For the USA, the FCC had split up the radio spectrum for business, commercial, and public use. Frequencies have been set aside for specific radio services and amateur radio use.

3 GMRS Specifics

The first two chapters cover the history of GMRS and the basics of radio. GMRS has specific frequences split into 22 channels and 8 repeater channels. This chapter will cover the frequencies and other technical details.

3.1 FCC Part 95

GMRS rules and specifications are defined in FCC Part 95 Subpart E, which can be found on the FCC website: https://www.ecfr.gov/current/title-47/chapter-I/subchapter-D/part-95. It is beyond the scope of the book to cover the whole specification, but this chapter will cover some of the rules and specifications. Other chapters will cover different parts of Subpart E.

Also note, FRS rules and specifications are defined in FCC Part 95 Subpart B. Some of the information overlaps both services. It is good to get to know these sections of the regulation as there are details for both users and radio manufacturers.

3.2 Channels and Frequencies

FRS and GMRS share the same frequencies, but there are limits to FRS. The 22 channels are broken up into 3 groups. The following chart covers the channels, frequency, and limitations.

Channel	Frequency (MHz)	Bandwidth		Transmit Power Limit (Watts)	
		FRS	GMRS	FRS	GMRS
1	462.5625	12.5 kHz	20 kHz	2	5
2	462.5875	12.5 kHz	20 kHz	2	5
3	462.6125	12.5 kHz	20 kHz	2	5

Channel	Frequency (MHz)	Bandwidth		Transmit Power Limit (Watts)	
		FRS	GMRS	FRS	GMRS
4	462.6375	12.5 kHz	20 kHz	2	5
5	462.6625	12.5 kHz	20 kHz	2	5
6	462.6875	12.5 kHz	20 kHz	2	5
7	462.7125	12.5 kHz	20 kHz	2	5
8	467.5625	12.5 kHz	12.5 kHz	0.5	0.5
9	467.5875	12.5 kHz	12.5 kHz	0.5	0.5
10	467.6125	12.5 kHz	12.5 kHz	0.5	0.5
11	467.6375	12.5 kHz	12.5 kHz	0.5	0.5
12	467.6625	12.5 kHz	12.5 kHz	0.5	0.5
13	467.6875	12.5 kHz	12.5 kHz	0.5	0.5
14	467.7125	12.5 kHz	12.5 kHz	0.5	0.5
15	462.5500	12.5 kHz	20 kHz	2	50
16	462.5750	12.5 kHz	20 kHz	2	50
17	462.6000	12.5 kHz	20 kHz	2	50
18	462.6250	12.5 kHz	20 kHz	2	50
19	462.6500	12.5 kHz	20 kHz	2	50
20	462.6750	12.5 kHz	20 kHz	2	50
21	462.7000	12.5 kHz	20 kHz	2	50
22	462.7250	12.5 kHz	20 kHz	2	50

Channels 8-14 limit the power output to 0.5 watts. This allows for FRS and GMRS to communicate at the same power level. The lower the power the shorter the range the radio can reach. Most GMRS radios have the power limits programmed in so you don't have to worry about being compliant.

The following table covers the 8 GMRS repeater channels and frequencies.

Channel	TX Frequency (MHz)	RX Frequency (MHz)	Power Limit (Watts)	Bandwidth (kHz)
23 (R15)	467.5500	462.5500	50	20
24 (R16)	467.5750	462.5750	50	20
25 (R17)	467.6000	462.6000	50	20

Channel	TX Frequency (MHz)	RX Frequency (MHz)	Power Limit (Watts)	Bandwidth (kHz)
26 (R18)	467.6250	462.6250	50	20
27 (R19)	467.6500	462.6500	50	20
28 (R20)	467.6750	462.6750	50	20
29 (R21)	467.7000	462.7000	50	20
30 (R22)	467.7250	462.7250	50	20

Different radios present the channel numbers either as 23-30 or R15-R22. The 8 repeater channels use the same receive frequency as channels 15-22. When communicating on channels 1-22, the communication type is known as simplex.

Definition: Simplex – The radio transmits and receives on the same frequency

For GMRS repeaters and radios, the communication type is called half duplex.

Definition: Half duplex – the radio has a frequency for transmit and another for receive. Communication switches between the two. The radio cannot transmit and receive at the same time.

There is an advantage matching the GMRS repeater channels with channels 15-22. Some radios have a talk-around option. If two people are communicating on a repeater and they are getting closer together compared to the distance from the repeater, both can bypass the repeater by pressing a button on the radio to talk around the repeater and use the simplex frequencies. Taking the repeater out of the equation can create clearer communication.

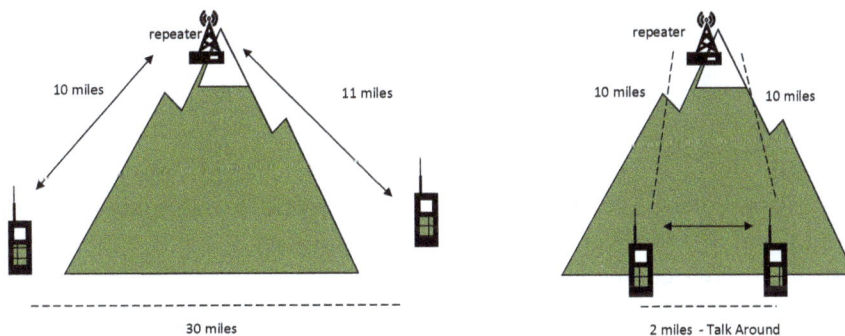

3.3   Line A and Line C

The FCC has rules about operating a GMRS radio near Canada. Specifically, Channel 19 (46.650 MHz) and Channel 21 (462.700 MHz) cannot be used North of Line A or East of Line C. Canada uses these frequencies for different operations. There is a coordination map found on the FCC website here: https://www.fcc.gov/reports-research/maps/frequency-coordination-canada/ and more details is found here: https://www.fcc.gov/engineering-technology/electromagnetic-compatibility-division/frequency-coordination-canada-below#LineA.

3.4   Squelch

GMRS radios have a built-in squelch feature. There is noise across the radio spectrum. Radios include a squelch circuity to suppress the noise and only pick up the stronger signal. Different radios offer different squelch levels to be set. The level can be adjusted to pick up stronger signals or turned off to pick up week signals.

3.5   Tone/Code Squelch (Not Privacy Tones)

With the wide use of cell phones, we are used to having direct private two-way communication. Even though you have a license to communicate on GMRS, you don't own the frequencies or channels. The channels are shared and everyone can hear what is being broadcasted on the channels. Since anyone can broadcast on the same channels, companies built on the squelch circuitry to add an additional tone coding scheme. Two people who have setup their radios for a specific code on a channel can talk to one another without interference by others who are using the same channel. The words privacy tones have been used to describe these tones, but "privacy" is not the right word. There is an example in the next section that will demonstrate the tone usage. There are two types of tone squelches that have been developed. One is based on an analog tone, and the other is a digital code.

3.5.1   CTCSS

Continuous Tone-Coded Squelch System (CTCSS) adds a low frequency tone to the transmitted signal. The receiver that has been configured to receive the tone and decodes the signal. If a signal comes in and doesn't have the tone, the signal / message is suppressed. The number of CTCSS tones has changed over the years. Here is the current list of CTCSS tones available.

CTCSS #	Tone (Hz)	CTCSS #	Tone (Hz)	CTCSS #	Tone (Hz)
1	62.5	19	118.8	37	183.5
2	67.0	20	123.0	38	186.2
3	69.3	21	127.3	39	189.9
4	71.9	22	131.8	40	192.8
5	74.4	23	136.5	41	196.6
6	77.0	24	141.3	42	199.5
7	79.7	25	146.2	43	203.5
8	82.5	26	150.0	44	206.5
9	85.4	27	151.4	45	210.7
10	88.5	28	156.7	46	218.1
11	91.5	29	159.8	47	225.7
12	94.8	30	162.2	48	229.1
13	97.4	31	165.5	49	233.6
14	100.0	32	167.9	50	241.8
15	103.5	33	171.3	51	250.3
16	107.2	34	173.8	52	254.1
17	110.9	35	177.3		
18	114.8	36	179.9		

3.5.2   DCS

Digital-Coded Squelch (DCS) was intended as a digital replacement for CTCSS. Rather than a low frequency tone, a digital bit steam with a 23-bit coded message. If a signal received has a matching code, the signal will continue on to be processed. If the signal doesn't have the matching code, the signal is squelched. Here is the current list of DCS codes:

DCS #	DCS Code	DCS #	DCS Code	DCS #	DCS Code
1	D023 (N or I)	29	D174 (N or I)	57	D445 (N or I)
2	D025 (N or I)	30	D205 (N or I)	58	D464 (N or I)
3	D026 (N or I)	31	D223 (N or I)	59	D465 (N or I)
4	D031 (N or I)	32	D226 (N or I)	60	D466 (N or I)
5	D032 (N or I)	33	D243 (N or I)	61	D503 (N or I)
6	D043 (N or I)	34	D244 (N or I)	62	D506 (N or I)
7	D047 (N or I)	35	D245 (N or I)	63	D516 (N or I)
8	D051 (N or I)	36	D251 (N or I)	64	D532 (N or I)

DCS #	DCS Code	DCS #	DCS Code	DCS #	DCS Code
9	D054 (N or I)	37	D261 (N or I)	65	D546 (N or I)
10	D065 (N or I)	38	D263 (N or I)	66	D565 (N or I)
11	D071 (N or I)	39	D265 (N or I)	67	D606 (N or I)
12	D072 (N or I)	40	D271 (N or I)	68	D612 (N or I)
13	D073 (N or I)	41	D306 (N or I)	69	D624 (N or I)
14	D074 (N or I)	42	D311 (N or I)	70	D627 (N or I)
15	D114 (N or I)	43	D315 (N or I)	71	D631 (N or I)
16	D115 (N or I)	44	D331 (N or I)	72	D632 (N or I)
17	D116 (N or I)	45	D343 (N or I)	73	D654 (N or I)
18	D125 (N or I)	46	D346 (N or I)	74	D662 (N or I)
19	D131 (N or I)	47	D351 (N or I)	75	D664 (N or I)
20	D132 (N or I)	48	D364 (N or I)	76	D703 (N or I)
21	D134 (N or I)	49	D365 (N or I)	77	D712 (N or I)
22	D143 (N or I)	50	D371 (N or I)	78	D723 (N or I)
23	D152 (N or I)	51	D411 (N or I)	79	D731 (N or I)
24	D155 (N or I)	52	D412 (N or I)	80	D732 (N or I)
25	D156 (N or I)	53	D413 (N or I)	81	D734 (N or I)
26	D162 (N or I)	54	D423 (N or I)	82	D743 (N or I)
27	D165 (N or I)	55	D431 (N or I)	83	D754 (N or I)
28	D172 (N or I)	56	D432 (N or I)		

Note: N (Normal) means the codes are positive. I (Inverted) mean the codes are negative. Radios can support both N and I codes, but in general practice GMRS sticks to N.

3.5.3 Tone/Code Implementation and Confusion

Different radio manufacturers implement the CTCSS tones and DCS codes different. The tones/codes are typically set for transmit and receive separately. Some manufactures have two separate lists for both tone types: TX-CTC, RX-CTC, TX-DCS, and RX-DCS, while others will combine the both tone list into one for transmit and receive: TX-CSS and RX-CSS. There is no standard so the names of the settings will be different.

Different radio manufactures have a tendency to pre-set the tones/codes in the radios. For example, let's say you purchase a GMRS radio from two different manufacturers. You set both radios to the same channel and attempt a transmission. If the radios cannot communicate with each other, it doesn't mean that they are incompatible. You have to check the squelch tones to make sure they match or turned off completely. The frequencies and tones are standard. GMRS radios from different manufactures can connected to one another so long as they are configured correctly. The next chapter covers configuring radios.

3.5.4 Example of Tone/Code Squelch (Not Privacy Codes) in Action Part1

There are some folks that still use the term Privacy Tones. Privacy is not the right word to use as the channels are open to everyone. For example, let's say we have three GMRS radios: A, B, and C. All three radios are set to channel 8. Radio A and B have the tones for RX and TX set to D032N. Radio C has the tones turned off.

Radio A
Channel 8
RX: CTS/DCS: D032N
TX: CTS/DCS: D032N

Radio B
Channel 8
RX: CTS/DCS: D032N
TX: CTS/DCS: D032N

Radio C
Channel 8
RX: CTS/DCS: OFF
TX: CTS/DCS: OFF

Here is what happens in this scenario:

- Radio A and B can communicate with each other.
- Radio C can hear the communication between A and B since the RX tones/codes are turned off.
- When Radio C transmits, neither Radio A or B can hear Radio C's transmission.

In this scenario, Radio A and B can have a conversation, but it is not "private" since Radio C can listen in. The only way Radio A or B can hear Radio C is by pressing the Monitor button. The Monitor button disables the squelch and tone/code squelch in order to listen for all transmissions on the channel.

3.6 A Few More Technical Details

Talk-around, monitor, squelch, and tone/code squelch are items that can be configured in a radio, which is leading to the next chapter. Before we move on, there are a few more details

3.6.1 Bandwidth: Narrow versus Wide

Bandwidth is called out on the channel table. The channel can be set to Wideband (25 kHz) or Narrowband (12.5 kHz). Channels 8-14 are fixed at Narrowband only. All other channels and the repeater channels can be adjusted to Narrowband or Wideband. The wider the band, the more data can be sent. Here is summary:

- Wideband
  - Better Audio Quality
  - Large amount of data that can be sent
  - Possible interference from other signals
  - High power consumption

- Narrowband
  - Reduce interference from other signals
  - Lower power consumption, which saves on battery.
  - Audio quality is not as great
  - Limited amount of data to be transmitted.

With batteries getting denser, the power consumption is not a problem. Most users prefer wideband to get the best sound quality.

3.6.2 TX Power

Transmission power can also be set in the radio. Channels 8-14 are locked to the lowest power of 0.5 Watts. All other channels including repeater channels can be set to low (0.5 Watts) or high 5 to 50 Watts depending on the radio. Most users set the power to high.

3.7 Transmission and Station Identification

Once you have a radio, you will want to start talking on it immediately. There are some rules that must be followed:

- For a single transmission, station identification needs to be broadcast; either by voice or morse code. Phonetic alphabet is recommended when using voice.
- For a continuous transmission / conversation, every 15 minutes, your station identification must be broadcast; either by voice or morse code. Phonetic alphabet is recommended when using voice.
- Advertisements and music cannot be transmitted.
- Be careful with your language as you don't know who is listening.

- If someone breaks into a conversation with an emergency, make sure to let that person provide information. If you can, relay the information to any emergency services.
- GMRS repeaters don't have to send station identification.

3.8   Summary

The important details of the GMRS specification have been covered. The frequencies, tones/codes, bandwidth, etc. are settings that can configured in a radio. The next two chapters look at specific radios and how they can be configured or modified.

4 GMRS Radios

The previous chapters covered the history, basics of radio, and specifications. Now, we will look into buying and configuring a GMRS radio.

Note: The manufacturers and websites, and radios discussed in this chapter are not an endorsement or paid advertising. The goal of the chapter is to provide information to make an informed radio purchasing decision.

4.1 The Right Radio for You

GMRS has grown significantly over the last 10 years. There are many GMRS radio makes and models. Each radio has a different set of features, but they all follow the spec so they can communicate with one another. Choosing the right radio depends on different factors. The information in this section is to help give you some idea of what to look for.

4.1.1 What are you going to use it for?

Chapter 1 discussed different uses for the GMRS radio. If all you want is something for emergencies, then a simple radio will suffice. If you do a lot of hiking then a radio with weather station, FM radio, and GPS might be the best fit. If you are camping with the family, then you might want a simple radio for the kids and a mobile radio for the base camp. You need to answer the question of what the radio is going to be used for. The answer might be a couple of radios for different situations.

4.1.2 Radio Types

To make choosing a radio even more exciting, there are three different types of radios to meet the different use scenarios.

- Handheld or walkie talkies – The most common type of radio. As the name implies these radios are held in your hand or clipped to your body. Since the antenna is close to you, the maximum power is 5 Watts. The lower power means they can handle all 22 channels and the 8 repeater channels. Some handhelds are programmable with memory to store different tone/code combinations for different channels.

- Mobile radio – This radio is put into a vehicle and the antenna is mounted on the outside roof. Mobile radios have a power output from 10 to 50 Watts, and can transmit over a long distance. Power comes from the vehicle battery. A mobile radio in a vehicle is ideal when communicate on long road trips or communicating with others going up a jeep trail. A mobile radio can also be used as a base station. For example, you want to have a higher power radio in the house to reach family members in the local area or the local repeater. Transmit on low power channels 8-14 can be limited or not available for a high transmit power mobile radio.

- Repeater – GMRS repeaters power output can range from 10 to 50 Watts. Repeaters come as a rack mounted for installation in a radio tower shack or as a mobile repeater that can be set up for local farms or emergencies. Repeater will be covered in the next chapter.

4.1.3   Features to Look For

Manufacturer put all types of features into their radios. The radio features can factor into the radio select decision. Here are some features to look for:

Feature	Comments
Battery with USB-C charging	A battery with more milliamp hours (mAh) the long the radio can be used. Most GMRS radio come with a docking cradle to do the charging. The cradle is bit outdated and impractical when all cellphones are being charged over USB-C cables. USB-C port and charging from a USB-C cable is must have feature.
FM radio receiver	Some radios allow for receiving FM broadcasts on VHF. You can listen to music or news. If a GMRS

	transmission is coming in, the radio can switch off the FM radio so you can listen to the transmission.
NOAA/Weather	The national weather broadcasts are also on VHF. Getting the weather report on a long camping trip can be useful. The
Compass / GPS	Ideal for hiking. Here are a few examples: BTECH GMRS-PRO, Wouxun KG-Q10G, and Garmin Rino 750t.
Data transmission	FCC part 95 allows for GMRS radios to send text / data in short bursts. The only radio that support sending text messages is the BTECH GMRS-PRO
VOX	Voice operated transmission - This allows you to use the radio hands free. Using an attached microphone, your voice can activate the transmit so you can continue to keep your hands on the wheel and eye on the road.
Case Rating	Working and playing outdoors can get a little rough. Keeping the radio protected from dust and moisture. The radios will list an International Protection Code (IP), here are some examples: IP55 – Some dust can get in, but will not interfere with the equipment. Can get sprayed with water with minimal effect. IP66 – Dust is blocked from getting in. Can get sprayed with water with minimal effect. IP67 – Dust and water are blocked. The device can survive being submerged into water to about 1 meter (3 ft 3 in).
Lamp	Some radios come with a lamp to act as a flash light.
Extra Channels	GMRS supports 22 channels and 8 repeater channels. Take into consideration a road trip where you might hit different repeaters and some of these

	repeaters use the same channel but different tone /codes, or you want to use a channel configured for a different set of tone / code squelch combos, have extra channels or memory slots to program is very helpful.
Programming software / Cable / Bluetooth	Speaking of setting up extra channels, many of the GMRS radios are programmable. The radios can be programmed using the onboard menu, using a software program is easier. You will want to check to see if the radio comes with programming software.

4.1.4 Programming Software and Cable

Most GMRS radio manufactures provide programming software for their radios. Some provide firmware upgrade tools. Cable is needed to program the radio. Sometimes the cable is included and sometimes it is sold separately. Check before buying the radio if the programming cable comes with the radio. The programming cable is typically a USB to Kenwood (K-Plug) connector:

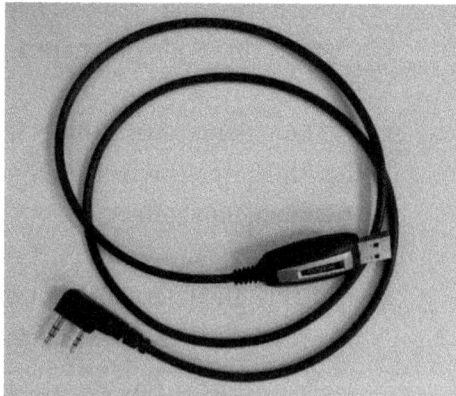

The programs from the manufacture are written for Windows. If you use MAC or Linux system, then you will have to add a Windows virtual machine to use them, or you can use an open-source tool called Chirp: https://chirpmyradio.com.

There are some drawbacks with CHIRP. CHIRP is community driven project, thus the radios supported are coming from the community. The support radio list is growing, but it might not have the latest radios from a manufacture. Also, the CHIRP might not support all the feature available to set like you would have available if you used the manufacture's software.

4.1.5   NOAA Stations

The National Weather Service broadcasts weather information for an area in the VHF band. The USA uses 7 frequencies, but GMRS radios will support 3 more channels for Canada and an extra for ASM2. The station listing website: https://www.weather.gov/nwr/station_listing. Some radios list the channel and some list the channel and frequency. Below is an example of a Channel / Frequency list:

Channel	Frequency (MHz)	Notes
CH1	162.550	
CH2	162.400	
CH3	162.475	
CH4	162.425	
CH5	162.450	
CH6	162.500	
CH7	162.525	
CH8	161.650	Canada (CMB)
CH9	161.775	Canada (CMB)
CH10	161.750	Canada (CMB)
CH11	162.000	ASM2 (Application Specific Message)

4.1.6 UHF Radio versus GMRS Radio

If you have a HAM license and a GMRS license, you can transmit on 420.0 MHz-450.0 MHz and all the GMRS frequencies at 462/467 MHz. Wouldn't it be great to have one radio access all those frequencies? Radio manufactures are thinking along those lines. Instead of developing 2 different radios, they develop one radio that supports 400 MHz – 470 MHz, and sell them under two different part-numbers. One is for GMRS and the other is sold as a UHF radio. The difference is in the tuning and firmware. FCC Part 95 sections 95.1771, 95.1773, 95.1775, and 95.1779 are very specific in how GMRS radio transmissions and unwanted transmissions are to be followed.

A few years ago, the Baofeng UV-5R brought the issue to the forefront. The UV-5R is a cheap and pretty decent HAM radio. It could be unlocked to support other frequencies including GMRS. It is legally permissible to listen to GMRS frequencies with the UV-5R, but a transmission would be considered illegal. Eventually Baofeng produced the UV-5G for GMRS.

The authors and publishers are not lawyers so the best answer is to the question: make sure that the radio is for GMRS and not the UHF until the FCC provides clear answer.

4.1.7 FCC ID Search

There is a way to check how the radio was licensed by the FCC using the FCC ID search. Once you have purchased the radio, you can get more information about the radio on the FCC website. Stamped on each radio is the model, serial number, and FCC ID of the radio. Here the stamp on the back of the Wouxun KG-805G (inside battery compartment).

1. Go to the https://www.fcc.gov/oet/ea/fccid website

2. Enter the first three letters of the FCC ID in he first box, and the rest of the FCC ID in the second box.

3. Hit the Search button. A new page appears with a records list matching the FCC ID.

4. Click on the summary and detail links to see the list of document and photos submitted to the FCC for approval.

You will be able to see if the radio was submitted as a GMRS radio or UHF radio.

4.1.8   FRS versus GMRS

As you search for GMRS radios, you will come across radios that appear to be GMRS radios, but are really FRS radios. In the latest change to FCC Part 95, radio manufacturers cannot make a hybrid FRS/GMRS radios. The radios have to be clearly marked as FRS or GMRS. FRS radios are limited to 2 Watts maximum transmission power. FRS radios are also limited to channels 1-22 so they cannot connect to repeaters.

4.1.9 Watch Out for Marketing

Companies do their best to sell their products. Radio manufacturers are no different so you need to be cautious of deceptive advertising. There are a couple marketing games to watch out for.

The first is the distance game. You might see a radio advertised as having the longest range. Being able to talk over long distances is great, but as discussed in chapter 2, UHF radios require clear line of site to talk over a great distance. Trees, terrain, and builds can block the signal and shorten the distance the radios transmission can reach.

The second is the channels game. You might see a GMRS radio advertises with 100 channels or over 3000 channel options. There are only 22 channels and 8 repeater channels. Add all the tone / code squelch and the possible combination are in the 1000s, but there are all using the 22 frequencies available. A better metric is the amount of "memory" or "programmable" channels are available. As mentioned earlier, a radio can be programmed with more channels by defining channels with different tone / code squelch values. For example, maybe your family has all their camping radios set to channel 15 with a specific tone / code pair. You decide to go off roading with some friends and they want to use channel 15 with different tone / code combination. You add a channel 31 that uses the frequency for channel 15 with the new tone / code combination.

There are some manufacturers that pre-program extra channels with tone / code squelch values. Pre-programming is fine. For example, you are going out hunting with some friends. If everyone is using the same radios with the pre-programmed frequencies, everyone can select a channel. No hunter wants to be in dear blind and have to break out a laptop to program the radio to a different tone / code squelch value.

It goes without saying, bypass all the marketing fluff and dig into the radio specifications and online user manuals.

4.2 The GMRS Radio Market

With the radio types, features, and what to watch out for covered, this section cover the radio manufactures, reviewers, and online stores.

4.2.1 GMRS Radio Manufacturers

The table below lists the known GMRS radio manufactures at the time of this writing. FRS radio manufactures are not on the list. Some of these manufactures are based in the USA, and some are based overseas. All radios are made in either China or Philippines. Some manufactures also make amateur, FRS, MURS, and other radios.

Manufacture	Website	Comments
Baofeng	baofengradio.com	
Bridgecom Systems	bridgecomsystems.com	GMRS Repeaters
BTECH	baofengtech.com	License agreement with Baofeng
Cobra	cobra.com	
Garmin	garmin.com	The only GMRS radio is Rino® 700 Series
Hytera	hytera.us	GMRS Repeaters
Midland	midlandusa.com	
Radioddity	radioddity.com	
Retevis	retevis.com	
Rugged	ruggedradios.com	
Wouxun	wouxun.com	

4.2.2 Radio Reviews

How can you try before buying a radio? Radio Shack has gone online. There are still specialty radio stores, but they are a little hard to find. The next best thing is to get a review. Besides online store review comments, here are a few YouTube channels and forums:

Channel	YouTube.com	Website	Comment
NotaRubicon Productions	@TheNotaRubicon	www.notarubicon.com	Leader of GMRS radio reviews
Ham Radio 2.0	@HamRadio2	www.livefromthehamshack.tv	Reviews radios for all radios services.
TheSmokinApe Ham Radio	@TheSmokinApe		Amateur radio information,

			radio review and antenna theory
myGMRS	@myGMRS	mygmrs.com	The forum page is a good place to ask and get information on GMRS radios
GMRS _two_way_radio	@GMRStwowayradio	gmrstwowayradio.com	Covers GMRS radios and other GMRS related topics

The list above is not extensive as there are many more online radio reviewers. The radios covered later in this chapter were purchased by the author. There are many more GMRS radios on the market, but the author misplaced the winning lottery ticket to buy all of them. Checking out online reviews is your best bet to get someone's insight on particular radio.

4.2.3 Where to Buy
Radios can be purchased directly from the manufacture's website. Here are a few specialty radio stores that specialize in GMRS radios. HAM Radio Outlet has physical stores across the USA.

Online Store	Website
Buy Two Way Radios	www.buytwowayradios.com
DX Engineering	www.dxengineering.com
HAM Radio Outlet	www.hamradio.com
myGMRS	shop.mygmrs.com/
R&L Electronics	www2.randl.com

4.3 GMRS Radio Technical Insights
The radios discussed in the following sections were paid for by the author. The goal of these sections is to provide the little technical details and tips that are not covered in the user manuals. Please refer to your radio's user manual for complete description and information.

4.4   Wouxun KG-805G

The Wouxun KG-805G is a very basic GMRS radio. You will not find the KG-805G listed on the Wouxun website. The radio was custom created for and sold by Buy Two Way Radios.

4.4.1   Features and Tips

The core features are:

- GMRS Channels 1-22
- GMRS Repeater Channels 23-30
- 128 programmable memory channels (30 standard GMRS and 98 memory spaces available)
- FM Radio
- VOX
- Alarm function
- Software programmable and CHIRP compatible.

The original battery was 1700 mAh, which require a charging cradle. A new 2800 mAh battery with USB-C charging capability is now available. The radio can now be charged up with a cell phone charger and cable.

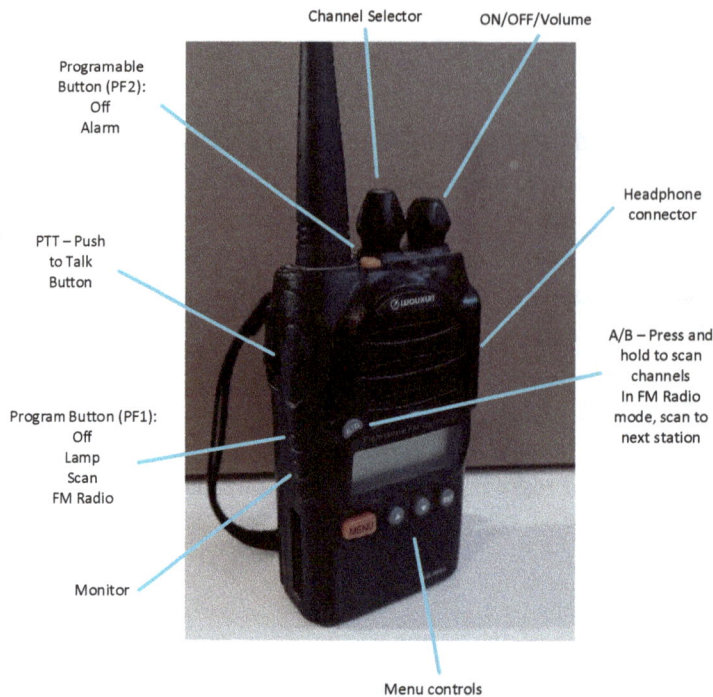

Channel Selector   ON/OFF/Volume

Programable
Button (PF2):
Off
Alarm

Headphone
connector

PTT – Push
to Talk
Button

A/B – Press and
hold to scan
channels
In FM Radio
mode, scan to
next station

Program Button (PF1):
Off
Lamp
Scan
FM Radio

Monitor

Menu controls

Since Buy Two Way Radios is a USA company, the manual for the radio is written in clear English. There are two buttons on the radio that can be programmed to a specific function. PF1 can turn on the LAMP, start a channel scan, or turn on/off the FM Radio. PF2 can be set for the Alarm. There is an A/B button on the unit, which the documentation is a little light on, but here is what was found in test:

- GMRS transmission, FM Radio is off: Tap A/B button turn on/off the LCD backlight
- GMRS transmission, FM Radio is off: Press and hold the A/B button to start the scan. Depending on the scan setting the scan will stop pause on a channel that is broad casting.
- FM Radio is on: Tap the A/B button and the tuner will scan for the next radio station broadcast.

When you turn on the FM radio, the radio's green light will flash as the radio tunes to the first radio station on the low end of the FM radio band. If a transmission comes in on the GMRS channel, the FM radio is silenced so the transmission can be heard. If after a few seconds there is no more transmission the KG-805G will switch back to FM radio.

The stock antenna does a good enough job, but the antenna can be replaced with a higher gain antenna. The antenna type is SMA-F. Buy Two Way Radios lists a compatible high gain whip antenna for the KG-805G: Melowave Bandit-G GMRS Handheld Radio Antenna.

4.4.2 Wouxon KG-805G Programming Software

The manual does a great job of listing all the menu items in the radio. Almost everything you will want to do with the radio can be performed in the radio menu. Adding more channels has to be completed using software.

1. Download KG-805G software from Buy Two Way Radios to your Windows computer. You can find the software in the download section on the KG-805G product page.
2. Unzip the file and run the installer.
3. Turn the radio on and set it to a channel that is not actively broadcasting, and then turn off the radio.
4. Connect the USB side of the programming cable to the Windows computer.
5. Open Device Manager by right click on Start button, and select Device Manager from the context menu.

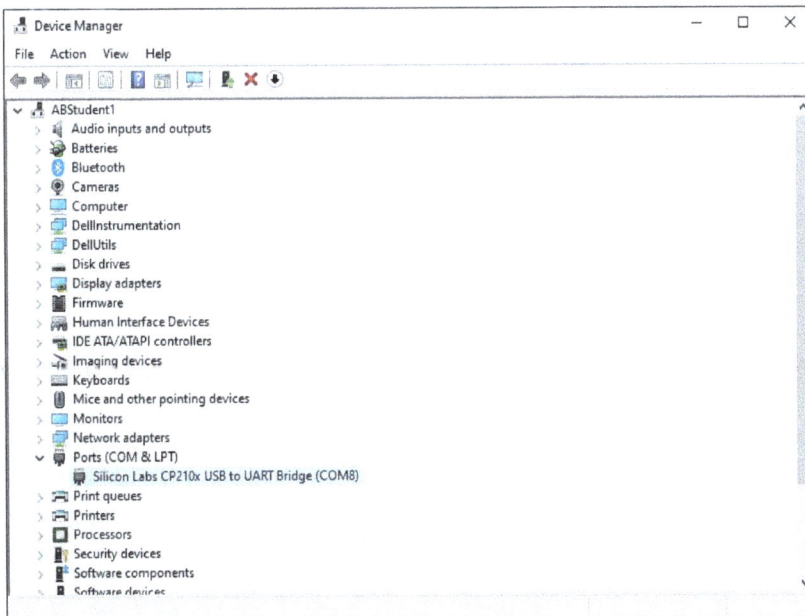

If you are using Windows 10 or Windows 11, the USB driver for the cable might already be installed. The USB cable should appear under ports. Different cable manufactures use different USB chips:

- Wouxun Programming Cable: Silicon Labs CP210X
- Retevis Programming Cable: Prolific USB-to-Serial

If the driver is not installed, you will have to install the driver from the CD provided with the programming cable or just run Windows update.

6. Plug the K-plug connector into the KG-805G
7. Turn on the radio
8. Start the KG-805G programming software.
9. From the menu, click on the "Communication Port".
10. In the dialog select the COM port that corresponds with the USB programming cable.

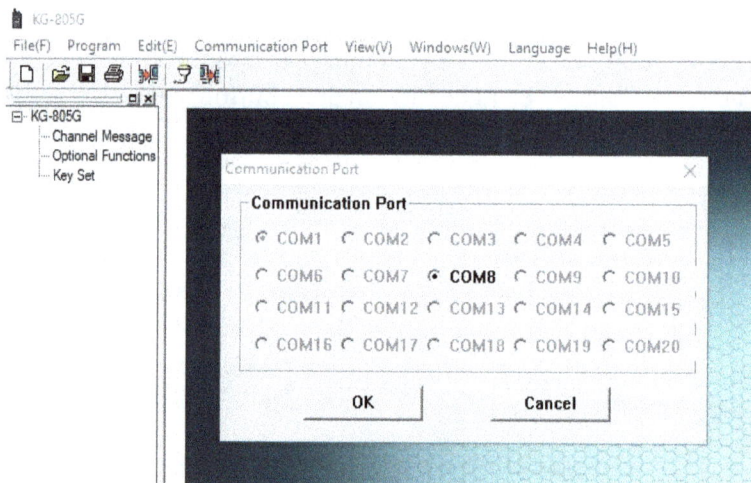

11. Either from the Program menu or the tool bar, click on "Read from Radio".

Note: There is a known issues that you may have to re-run step 9 as the first read doesn't pull down all the radio information.

12. Click on the OK button in the ReadSuccessed dialog.

F	OFF	High	Wide	OFF	QT
F	OFF	High	Wide	OFF	QT
F	OFF	Low	Narrow	OFF	QT
F	OFF	Low	Narrow	OFF	QT
F	KG-805G			×	QT
F					QT
F					QT
F		ReadSuccessed			QT
F					QT
F					QT
F		OK			QT
F					QT
F					QT
F	OFF	High	Wide	OFF	QT
F	OFF	High	Wide	OFF	QT
F	OFF	High	Wide	OFF	QT
F	OFF	High	Wide	OFF	QT

The Channel Message window will list all the GRMS channels, frequencies, Tones/Codes, and bandwidth settings. You will have to scroll to the right to see and set the channel names. Scroll down and you can see all 128 channels that can be programmed. Notice, the RX CTCSS/DCS and TX CRSS/DCS are separate columns. The radio menu breaks these out into 4 menu items.

13. In the left directory pane, click on the Option Functions to open the Optional Functions window.

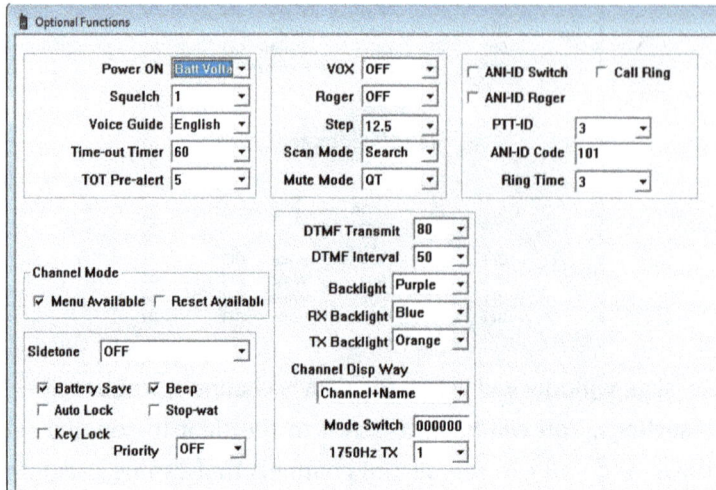

Many of the optional features can be set from the radio menu. The ANI and DTMF settings don't appear to do anything. Keep in mind that some software gets recycled for use with other radios.

14. In the left directory pane, click on the Key Set. This will open a window to set the programmable buttons.

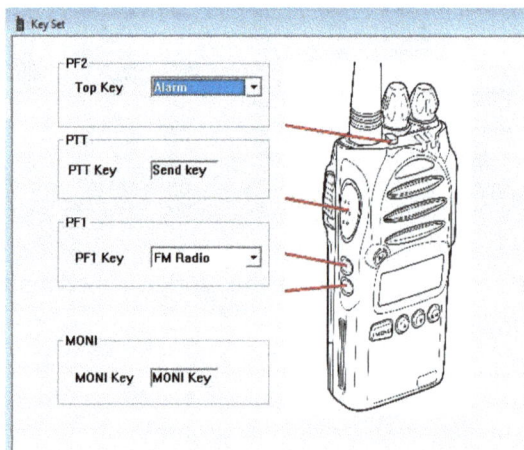

15. If you make any changes, you can apply the changes by select from the menu or the toolbar, Write to Radio. The radio will flash at the changes are programmed in.

16. When finished, close the program and turn off the radio.

17. Disconnect the radio from the K-plug. Turn on the radio, and you should see the changes that have been made.
18. Be sure to save the settings to a file.

4.4.3 Wouxon KG-805G Programming with CHIRP

Besides the Wouxon program software, there is also support for CHIRP. The KG-805G is support

1. Download CHIRP from chirp.danplanet.com to your computer.
2. Run the installer.
3. Turn the radio on and set it to a channel that is not actively broadcasting, and then turn off the radio.
4. Connect the USB side of the programming cable to the Windows computer.
5. Plug the K-plug connector into the KG-805G.
6. Turn the radio on.
7. Open the CHIRP software.
8. From the menu, select Radio-> "Download from radio…".
9. A dialog appears. Select the COM port, radio vendor, and the model. Click OK and the data from the radio is pulled in.

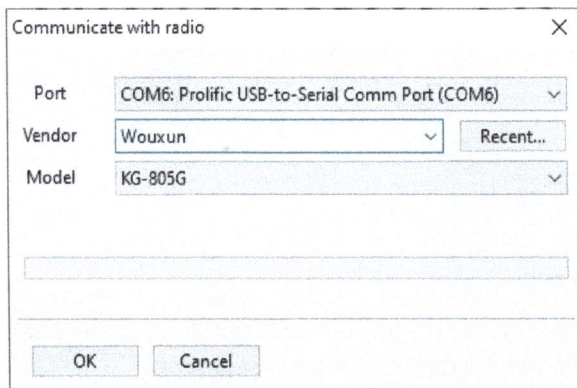

There are only two tabs for the radio settings. The first lists all the channels. You will see the Tone/Codes are presented differently. TSQL stands for tone squelch and DTCS is for DCS. Off set for repeater channels is shown, but this is standard so don't changes this setting. Keep in mind CHIPR was developed for HAM radios so managing the offset frequencies is a HAM thing.

	Frequency	Name	Tone Mode	Tone	Tone Squelch	DTCS	RX DTCS	DTCS Polarity	Duplex	Offset/ TX Freq	Cross Mode	Mode	Skip	Power	Comment
7	462.712500	CH07										FM		High	
8	467.562500	CH08	DTCS			032		NN				NFM		Low	
9	467.587500	CH09										NFM		Low	
10	467.612500	CH10										NFM		Low	
11	467.637500	CH11										NFM		Low	
12	467.662500	CH12										NFM		Low	
13	467.687500	CH13										NFM		Low	
14	467.712500	CH14										NFM		Low	
15	462.550000	CH15										FM		High	
16	462.575000	CH16										FM		High	
17	462.600000	CH17										FM		High	
18	462.625000	CH18										FM		High	
19	462.650000	CH19										FM		High	
20	462.675000	CH20										FM		High	
21	462.700000	CH21										FM		High	
22	462.725000	CH22										FM		High	
23	462.550000	HEMET	Cross			125	245	NN	+	5.000000	DTCS->DTCS	FM		High	
24	462.575000	PALMT	TSQL		162.2				+	5.000000		FM		High	
25	462.600000	RPT17							+	5.000000		FM		High	
26	462.625000	RPT18							+	5.000000		FM		High	
27	462.650000	KVGMRS	DTCS			077		NN	+	5.000000		FM		High	
28	462.675000	PCOMM	TSQL		162.2				+	5.000000		FM		High	
29	462.700000	RPT21							+	5.000000		FM		High	
30	462.725000	RPT22	DTCS			131		NN	+	5.000000		FM		High	
31															

Click on the Settings tab, and you will see only the basic radio settings that can be set. All the other settings and programable button options are not available. CHIRP provides access to change only the basic settings.

4.5 Retevis RT76P

The Retevis RT76P is a feature rich GMRS radio compared to the basic Wouxon KG-805G. Different type of radio to hit a different market segment.

4.5.1 Features and Tips
The Retevis RT76P is an older GMRS radio, but it is packed with features:

- GMRS Channels 1-22
- GMRS Repeater Channels 23-30
- 128 programmable memory channels (30 standard GMRS and 98 memory spaces available)

- Dual channel reception
- NOAA Stations CH1-CH11
- FM Radio Reception
- 2M (VHF) / 70cm (UHF) frequency reception (RX only)
- Keypad
- DTMF / ANI messaging
- Group calling
- Alarm function
- VOX
- Power on picture support
- Programmable with software, and CHIRP support. Programming cable is sold separately.

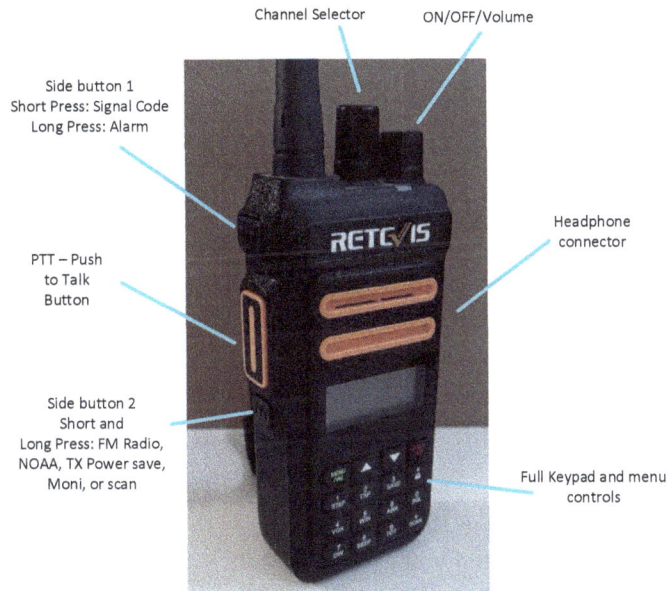

Early online reviewers point out that the manual has some mistakes. The Chinese to English translation is not 100% perfect. The biggest error is in the menu listing of the manual. The manual includes two additional items No. 26 Direction Menu No 27 Offset. These items are not used for GMRS and are not available in the unit. The RT76P is also a UHF radio, and these two options are import for HAM operators. All the other menu items are there.

The battery is 1400 mAh, which is good for 10 hours. The radio doesn't support direct USB-C charging. The USB to charge cradle is required.

The only programmable button on the radio is Side Button 2 (manual) or SKEY (program software). The button can be set for long press or short press to the following items: FM radio, TX Power (power save), Moni (monitor), Scan, and Weather.

The Menu / VM button has a dual purpose. Short press goes into the menu. Long press changes the to VFO or frequency mode. In VFO, mode you can scan and listen to VHF and UHF frequencies. The software allows you to preset the starting frequencies of VHF or UHF for each band.

The Exit / AB button has a dual purpose. Exit simple exits from the menu item. Long press switches the tuning of the A and B bands.

The Scan button on the key pad can scan channels in GMRS mode or scan frequencies in VFO or FM radio mode. The scan mode actions can be set on timer, stop briefly or just stop when a transmitting frequency/ channel is found.

Navigating the menu items can be performed using the channel knob, up / down arrows, or use the quick shortcuts on the keypad to access the first 10 popular items.

One of the biggest remarks was the reception on FM radio and NOAA channels. The physical fix was to get a gain antenna. The antenna on the RT76P has a screw holding antenna down, but the antenna can be unscrewed. Since the radio can receive VHF and UHF frequencies, the antenna needs to match. The antenna is a SMA-M antenna. Retevis offers their own branded SMA-M whip antenna. When installed, the stations come in load and clear. What really fixed the reception and other issues, was the firmware update. The next section discusses the firmware update.

4.5.2   Retevis RT76P Firmware Update
Any new orders for the RT76P should come with the latest firmware update already installed. If the firmware has been updated, the following should be fixed:

- Channel 8 was set to 467.5675 MHz, where it should be set at 467.5625 MHz.
- Menu Number 31 – PROGER versus ROGER. This was a typo in firmware and manual.
- The reviewers also pointed out that some menu items couldn't be changed on the device. The firmware update fixes these issues. DTMF and ANI are software only.
- FM and NOAA reception is better.

The only minor thing that is outstanding is menu number 5 for Bandwidth. The unit's menu says "width" instead of "wide".

If channel 8 is coming up with the wrong frequency, then you will have to update the firmware. The latest firmware will fix the short comings with the original firmware regarding menus, frequency settings, and NOAA stations. A Windows computer is needed to run the software. The firmware comes with the binary and an executable. The executable is called: RT76P_Bootloader固件升级工具.exe, which translates to RT76P_BootloaderFirmwareUpgradeTool.exe.

1. Download the latest firmware from the Retevis site to a Windows PC.
2. Extract the contents of the zip file.
3. Turn the radio on and set it to VFO mode, and then turn off the radio.
4. Connect the programming cable USB end to the computer USB port.

Note: Check device to make sure the device driver has been installed for the USB cable.

5. Connect the K-prong header of the programming cable to the RT76P head phone connector.
6. Turn the radio on.
7. Run the RT76P_Bootloader固件升级工具.exe
8. Click on the three dots (…) button and open the .binX file.
9. Select the COM port for the USB cable.
10. Click Upload button.

11. When the firmware has been updated, the radio will reboot. Close the firmware update program.

4.5.3   Retevis RT76P Programming Software

One pet peeve for first time buyers is that Retevis radios come with CTCSS/DCS codes already set, and another pet peeve is that the GMRS channels that support wide band and should be set to wide band are set to narrow band. The manual clearly points out that there are CTCSS/DCS defaults preconfigured in the unit. If these were set to nothing, the radio would be ready out of the box to communicate with other radios, but the codes are set to communicate with other Retevis radios. All the CTCSS/DCS codes can be configured in the menu, but it is much easier with the software.

The software to program the RT76P can be downloaded from the Retevis site. One reviewer pointed out that they download the programming software and tried to perform a read of the data from the radio, and the application crashed. There is a trick to the software. You have to run the program with Administrator rights. Here are the steps:

1. Download the latest RT76P programming software from the Retevis website.
2. Run the installer.
3. Once the installer has finished, right click on the RT76P.exe or shortcut.
4. Select properties
5. Click on the Compatibility tab.
6. Check the "Run this program as an administrator"
7. Click Apply
8. Click OK

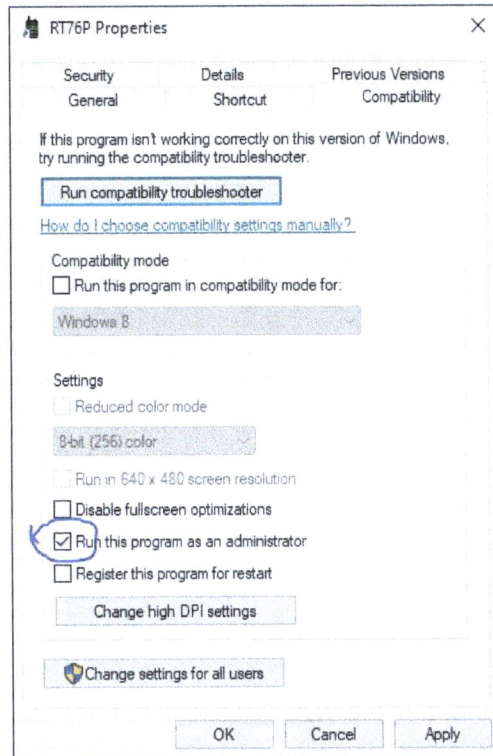

9. Turn the radio on and set it to VFO mode, and then turn off the radio.
10. Connect the programming cable USB end to the computer USB port

Note: Check device to make sure the device driver has been installed for the USB cable.

11. Connect the K-prong header of the programming cable to the RT76P head phone connector.
12. Turn the radio on.
13. Open the RT76P software.
14. From the menu, select Settings -> Port and set the COM port for the radio.
15. From the menu, select Program-> Read from Radio

Note: Reading and Writing data to the device will ask for a password. There is no password, and you can just click start.

Once the data from the radio has been filled in, you will see the list of channels. Scroll down and you will see all 128 channels that can be configured. The RX CTCSS/DCS and TX CTCSS/DCS are split into 4 menu items in the radio. In the software, there are split into two columns RX QT/DQT and TX QT/DQT. QT is an old Kenwood name for CTCSS. DQT or digital Squelch tone is for DCS. Why the program calls them differently is anyone's guess.

RT76P

File(F)   Edit(E)   Program(P)   Setting(S)   Help(H)

New Save Open Read Write

Channel Information

Channel	Rx Freq	Rx QT/DQT	Tx Freq	Tx QT/DQT	Power	W/N	PTT-ID	BCL	Scan	Signal	Compand	Name	ANI
3	462.61250	OFF	462.61250	OFF	M	W	OFF	OFF	ON	1	OFF		OFF
4	462.63750	OFF	462.63750	OFF	M	W	OFF	OFF	ON	1	OFF		OFF
5	462.66250	OFF	462.66250	OFF	M	W	OFF	OFF	ON	1	OFF		OFF
6	462.68750	OFF	462.68750	OFF	M	W	OFF	OFF	ON	1	OFF		OFF
7	462.71250	OFF	462.71250	OFF	M	W	OFF	OFF	ON	1	OFF		OFF
8	467.56250	D032N	467.56250	D031N	L	N	OFF	OFF	ON	1	OFF		OFF
9	467.58750	OFF	467.58750	OFF	L	N	OFF	OFF	ON	1	OFF		OFF
10	467.61250	OFF	467.61250	OFF	L	N	OFF	OFF	ON	1	OFF		OFF
11	467.63750	OFF	467.63750	OFF	L	N	OFF	OFF	ON	1	OFF		OFF
12	467.66250	OFF	467.66250	OFF	L	N	OFF	OFF	ON	1	OFF		OFF
13	467.68750	OFF	467.68750	OFF	L	N	OFF	OFF	ON	1	OFF		OFF
14	467.71250	OFF	467.71250	OFF	L	N	OFF	OFF	ON	1	OFF		OFF
15	462.55000	OFF	462.55000	OFF	M	W	OFF	OFF	ON	1	OFF		OFF
16	462.57500	OFF	462.57500	OFF	M	W	OFF	OFF	ON	1	OFF		OFF
17	462.60000	OFF	462.60000	OFF	M	W	OFF	OFF	ON	1	OFF		OFF
18	462.62500	OFF	462.62500	OFF	M	W	OFF	OFF	ON	1	OFF		OFF
19	462.65000	OFF	462.65000	OFF	M	W	OFF	OFF	ON	1	OFF		OFF
20	462.67500	OFF	462.67500	OFF	M	W	OFF	OFF	ON	1	OFF		OFF
21	462.70000	OFF	462.70000	OFF	M	W	OFF	OFF	ON	1	OFF		OFF
22	462.72500	OFF	462.72500	OFF	M	W	OFF	OFF	ON	1	OFF		OFF
23	462.55000	D245N	467.55000	D125N	M	W	OFF	OFF	ON	1	OFF	Nemet	OFF
24	462.57500	162.2	467.57500	162.2	M	W	OFF	OFF	ON	1	OFF	Palomar	OFF
25	462.60000	OFF	467.60000	OFF	M	W	OFF	OFF	ON	1	OFF		OFF

COM6   VHF:136-174MHz   Version v1.5.5

From the Edit menu you can change the Optional Features, DTMF tones, and upload a picture to appear on the LCD on boot up. The option window sets all the radio options for VOX, VFO A/B band settings, scan settings, Skey button long and short press, etc.

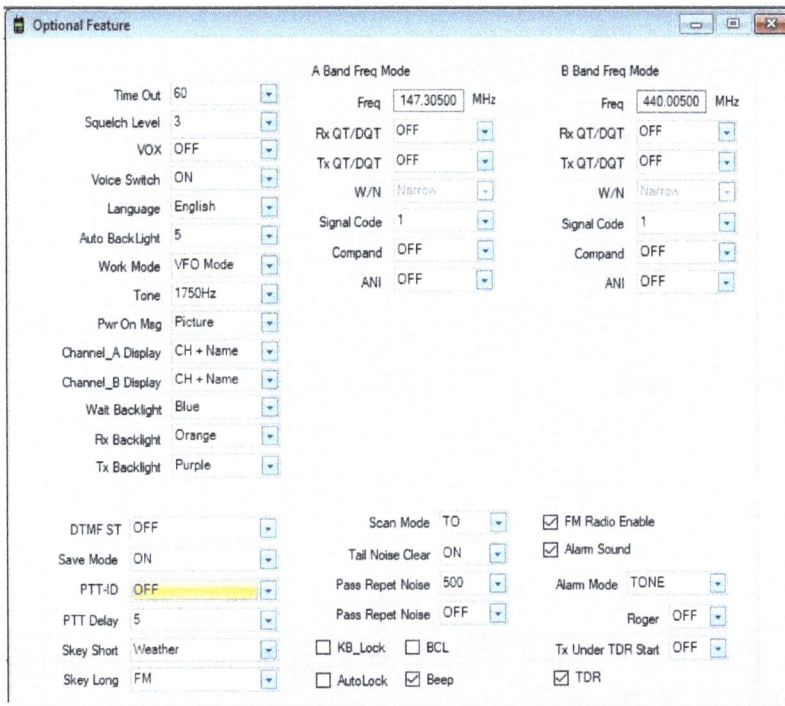

Note: If you set the Channel_A Display (MDF-A) or Channel_B Display (MDF-B) options are set to show the channel + name, the display will show the frequency if a channel name has not been set.

The DTMF window is used to for group messaging. The page also contains the custom Automatic Number ID (ANI-ID) code which can be sent when you transmit.

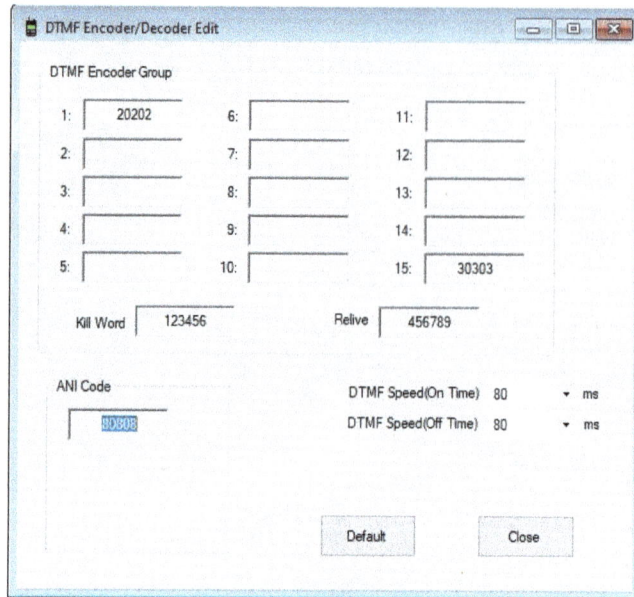

The last feature is either to have a power on image or message. A power on logo or message allows for personalization or define a company radio.

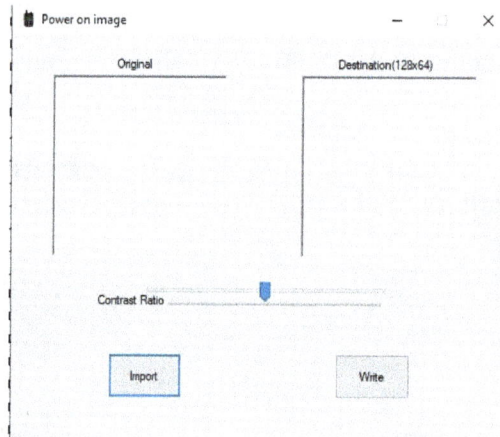

16. Once you have made all the changes, from the menu select Program -> Write to radio... or the write to radio on the toolbar.
17. A dialog appears asking to set password, just ignore this and hit start. The settings will be stored in the radio.
18. Be sure to save the settings to a file.

4.5.4   Retevis RT76P Programming with CHIRP

CHIRP also supports the RT76P.

1. Download CHIRP from chirp.danplanet.com to your computer.
2. Run the installer.
3. Turn the radio on and set it to frequency mode outside the GMRS frequencies, and then turn off the radio.
4. Connect the programming cable USB end to the computer USB port
5. Connect the two-prong header of the programming cable to the RT76P head phone connector.
6. Turn the radio on.
7. Open the CHIRP software.
8. CHIRP will allow you to immediately read the data from the device. Select the port, radio vendor, and the model. Click OK and the data from the radio is pulled in.

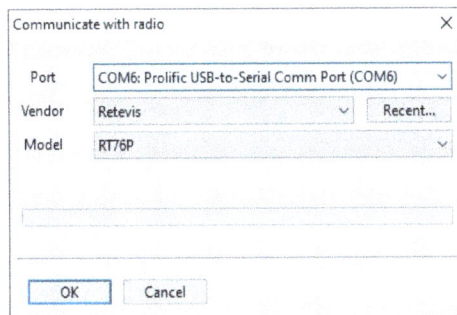

The listing of the memory channels is displayed. The proper CTCSS/DCS names are used.

One the settings tab, you will find the CHIRP has most of the available settings found in the Retevis program software.

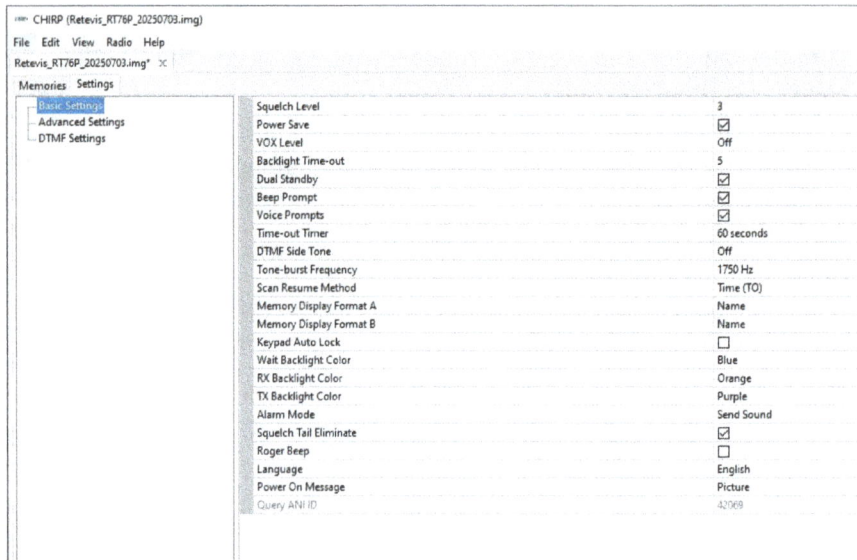

CHIRP (Retevis_RT76P_20250703.img)		
File Edit View Radio Help		
Retevis_RT76P_20250703.img* ×		
Memories **Settings**		
Basic Settings	Squelch Level	3
Advanced Settings	Power Save	☑
DTMF Settings	VOX Level	Off
	Backlight Time-out	5
	Dual Standby	☑
	Beep Prompt	☑
	Voice Prompts	☑
	Time-out Timer	60 seconds
	DTMF Side Tone	Off
	Tone-burst Frequency	1750 Hz
	Scan Resume Method	Time (TO)
	Memory Display Format A	Name
	Memory Display Format B	Name
	Keypad Auto Lock	☐
	Wait Backlight Color	Blue
	RX Backlight Color	Orange
	TX Backlight Color	Purple
	Alarm Mode	Send Sound
	Squelch Tail Eliminate	☑
	Roger Beep	☐
	Language	English
	Power On Message	Picture
	Query ANI ID	42069

The Advanced settings have the Alam and side key settings. The DTMF page has all the DTMF and ANI settings.

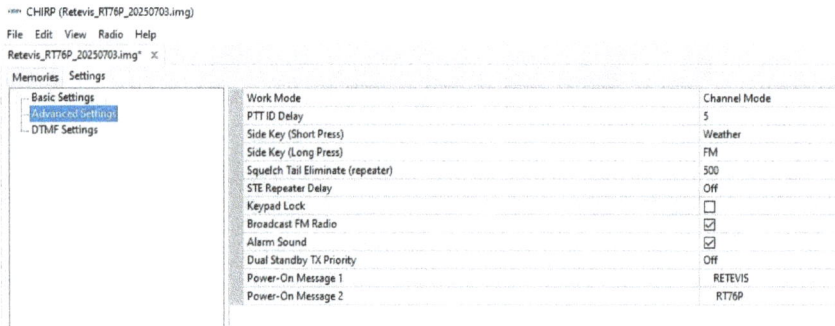

CHIRP (Retevis_RT76P_20250703.img)		
File Edit View Radio Help		
Retevis_RT76P_20250703.img* ×		
Memories **Settings**		
Basic Settings	Work Mode	Channel Mode
Advanced Settings	PTT ID Delay	5
DTMF Settings	Side Key (Short Press)	Weather
	Side Key (Long Press)	FM
	Squelch Tail Eliminate (repeater)	500
	STE Repeater Delay	Off
	Keypad Lock	☐
	Broadcast FM Radio	☑
	Alarm Sound	☑
	Dual Standby TX Priority	Off
	Power-On Message 1	RETEVIS
	Power-On Message 2	RT76P

The Settings tab contains the same Basic, Option and DTMF settings that are found in the Retevis software.

4.6   Retevis C2

The Retevis C2 is successor replacement for the RT76P. The receiver has better sensitivity and doesn't require a whip gain antenna to tune in the NOAA and FM frequencies.

4.6.1 Features and Tips

The Retevis C2 is has the following features:

- GMRS Channels 1-22
- GMRS Repeater Channels 23-30
- 256 programmable memory channels (30 standard GMRS and 226 memory spaces available)
- Dual channel reception
- AI Noise reduction
- NOAA Stations CH1-CH11
- FM Radio Reception
- Keypad
- DTMF / ANI messaging
- Group calling
- Alarm Local and Remote
- VOX
- Radio pairing
- Flash light
- USB-C charging port
- Programmable with software. Programming cable is sold separately.

Flashlight
ON/OFF/Volume
Headphone connector
PTT – Push to Talk Button
Channel / Menu Selector
Side Key Programable
USB-C Charing Port
Menu Exit / Band Select
Short Press: Menu / Long Press: Record
P2 Programable
P1 Programable
Full Keypad and menu controls

The C2 has some clear physical differences from the RT76P. The C2 has a color screen and flash light on top. The channel knob / menu select has been replaced with an up-down rocker button on the keypad. The menu and exit buttons are labelled a little differently.

The biggest difference is the 3 programable buttons: P1, P2, and Side Key (Skey). The short and long press of these keys can be set to one of the following: OFF, Local alarm, Remote alarm, Scan, Noise reduction (AI DNS <Denoise>), VOX, NOAA, TX power, Monitor, Flashlight, Talk around, Reverse frequency, and Detect Frequency. The Side Key (Skey) has a few more options with a short press: Group call PTT and SUB-PTT.

The menu system is a little different. The main menu that is split up into 5 items. The manual does a good job of covering all the menu items. Unfortunately, the first menu item is called "Intercom Settings", which should be "Radio Settings" per the manual. The manual has a typo on page 5, where the title above the table should read "Radio Settings" not "Icon Instructions". The Radio settings cover LCD brightness, timeouts, power saving, etc. The 2nd item is "Channel Settings", which is for the active main selected channel. The Channel Settings cover squelch, tone / code squelch, bandwidth, channel name, etc. The 3rd item is for "Audio" settings: AI noise reduction, Recorder, Tone beep, Antenna Mic gain, etc. The 4th item is for the DTMF functions. The 5th item is for configuring the programable buttons.

4.6.2   Channel Synchronization

The C2 is advertised with the idea to synchronize a single channel's configuration with other radios. The radios can be any GMRS radio from any manufacturer. How this works:

1. Set a different GMRS radio channel 7 with RX/TX DCS to D023N. This is just an example. The channel and tone / codes can be anything.
2. In the C2, configure the Skey short press to detect frequency
3. In the C2, set the main active channel to channel 7.
4. Tap on the Skey to start the detect frequency. The function can detect VHF and UHF so use the up-down button to make sure UHF is selected.
5. Key up (hit the PTT button) on the other radio. Hold the PTT button for a few seconds, and the C2 should show the frequency and CSS value.

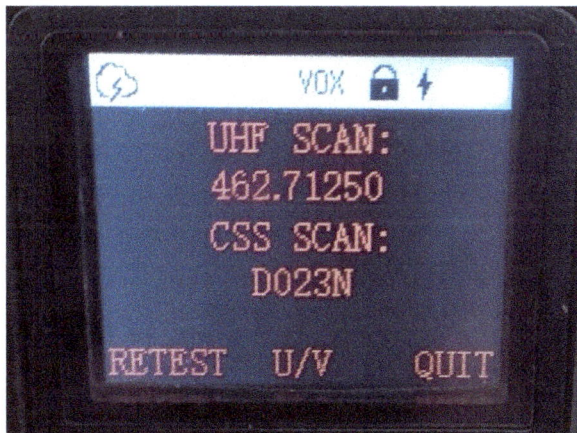

Note: You may have to retest a few times for the function to work correctly.

6. Tap on the Skey to save the settings for channel 7.
7. Go to the channel settings for Channel 7, and you will see that the DCS will be configured to D023N. This was only for one channel, you have to repeat steps 4-6 for a different channel.

Looking at the DCS on the screen, you will notice a *:SW in the lower left corner. All the menu items in the radio can be modified without the programming software. Except for the CTCSS/DCS settings. If the CTCSS/DCS settings are turned off, they cannot be enabled and set within the menu. Once enabled, they can be changed to any other code, but they can only be disabled using the software. The inability to have full control from the menu could be considered a draw back for the C2.

4.6.3   Retevis C2 Programming Software

CHIRP doesn't support the C2 as of this writing. Like the RT76P, the C2 comes with tone / code squelch preset on channels 1-22. The manual lists what each channel is set too. The software has to be used to clear out the preset tone / code squelch for each channel. The software to program the C2 can be downloaded from the Retevis site.

1. Download the latest C2 programming software from the Retevis website.
2. Run the installer.
3. Turn the radio on and set it to a channel that is not actively broadcasting, and then turn off the radio.
4. Connect the USB side of the programming cable to the Windows computer.
5. Open Device Manager by right click on Start button, and select Device Manager from the context menu.

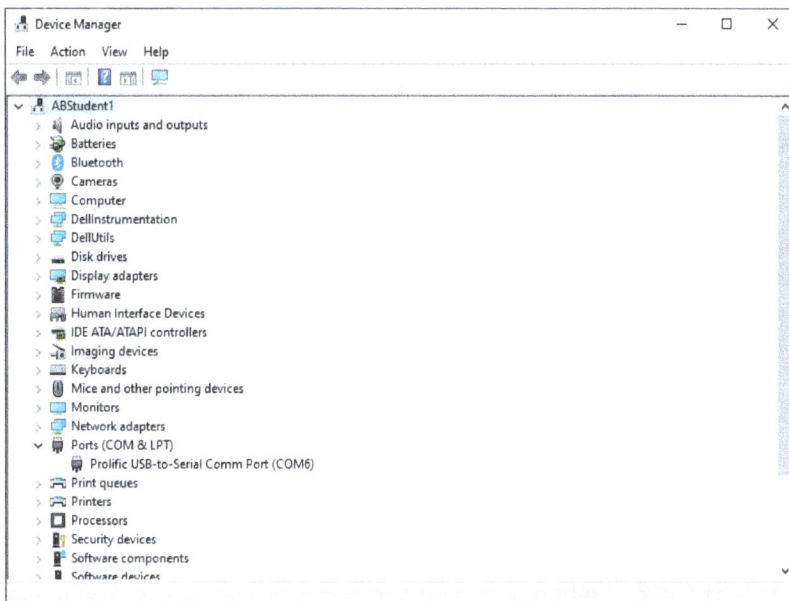

If you are using Windows 10 or Windows 11, the USB driver for the cable might already be installed. The USB cable should appear under ports. Different cable manufactures use different USB chips:

- Wouxun Programming Cable: Silicon Labs CP210X
- Retevis Programming Cable: Prolific USB-to-Serial

If the driver is not installed, you will have to install the driver from the CD provided with the programming cable or just run Windows update.

6. Plug the K-plug connector into the C2
7. Turn on the radio
8. Start the C2 programming software.
9. From the menu, click on Settings -> Set Com.
10. In the dialog select the COM port that corresponds with the USB programming cable, and click OK.
11. From the menu, select Program-> Read from Radio.

Note: Reading and Writing data to the device will ask for a password. There is no password, and you can just click start.

12. Once the read finishes, click on Cancel to close the dialog.

Once the data from the radio has been filled in, you will see the list of channels. Scroll down and you will see all 256 channels that can be configured. The menu on the left opens other windows to the different radio settings. The CTCSS/DCS is correctly labeled. The lower power channels 8-14 are marked with grey as these settings a cannot be changed.

Click on the More >>for one of the channels, and a dialog with the channel's settings will appear.

The Function page contains all the radio and programmable button settings.

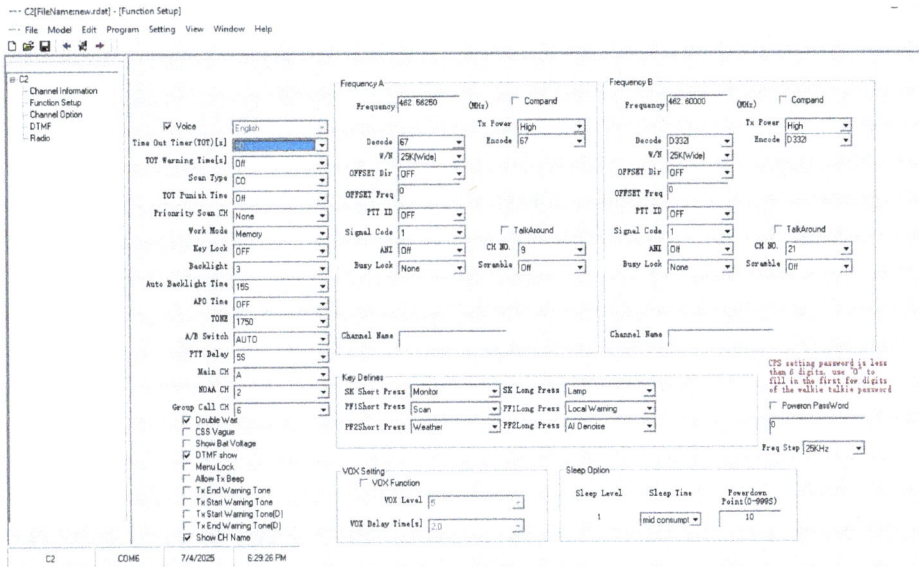

Channel options as the squelch level and AI DNS (Denoise) level settings.

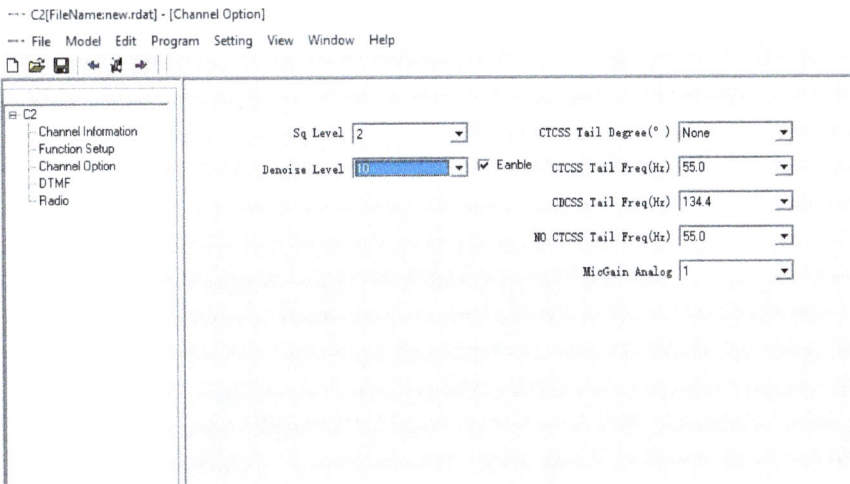

DTMF contains the ANI (Local ID) settings and DTMF codes.

--- C2[FileName:new.rdat] - [DTMF]

--- File Model Edit Program Setting View Window Help

Dtmf Code

NO.	Code
1	
2	
3	
4	
5	
6	
7	
8	
9	
10	
11	
12	
13	
14	
15	
16	

Dtmf Setting

Local ID
Separate Code *
Group Code A
Model Status Normal

Auto Reset Time(S) 7
First Code Time(ms) 200

☑ Side Tone

PTT ID(BOT)
PTT ID(EOT)
Stun Code
Die Code
Wakup Code

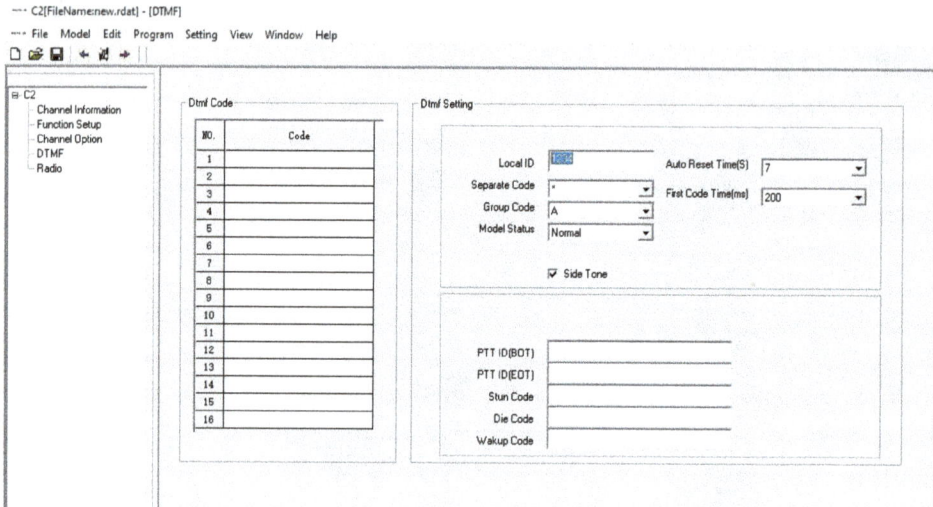

The Radio window allows you to preset FM radio station frequencies that the channel selector button can jump too.

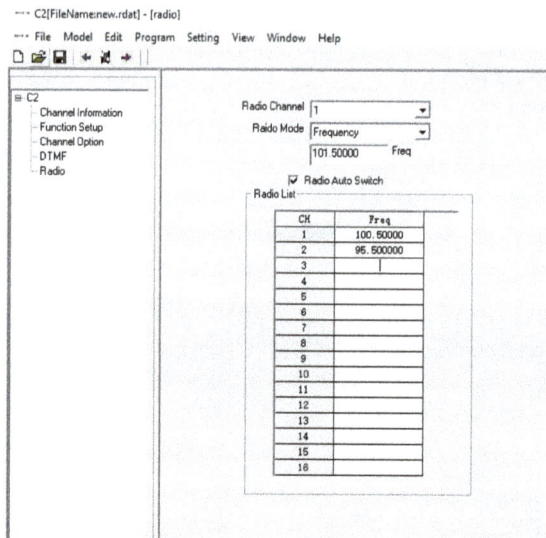

--- C2[FileName:new.rdat] - [radio]

--- File Model Edit Program Setting View Window Help

- C2
  - Channel Information
  - Function Setup
  - Channel Option
  - DTMF
  - Radio

Radio Channel 1
Raido Mode Frequency
101.50000 Freq

☑ Radio Auto Switch

Radio List

CH	Freq
1	100.50000
2	95.500000
3	
4	
5	
6	
7	
8	
9	
10	
11	
12	
13	
14	
15	
16	

13. Once you have made all the changes, from the menu select Program -> "Write to radio..." or the write to radio button on the toolbar.
14. A dialog appears asking to set password, just ignore this and hit start. The settings will be stored in the radio.
15. Be sure to save the settings to a file.

4.7 Retevis RA86

The Retevis RA86 is a small mobile radio that can be installed in to a vehicle or used as a base-station. All the radio control buttons are on the handheld microphone.

4.7.1 Features and Tips

The RA86 is a very simple GMRS Radio. Because of its power, channels 8-14 are set for receive only.

- GMRS Channels 1-7 (RX / Tx power 5W max)
- GMRS Channels 15-22 (RX / Tx power 15 max)
- GMRS Channel 8-14 (RX only)
- GMRS Repeater Channels r15-r22
- 99 programmable memory channels (30 standard GMRS and 66 memory spaces available)
- NOAA Stations CH1-CH10
- Vox
- Programmable with software. Programming cable is sold separately.

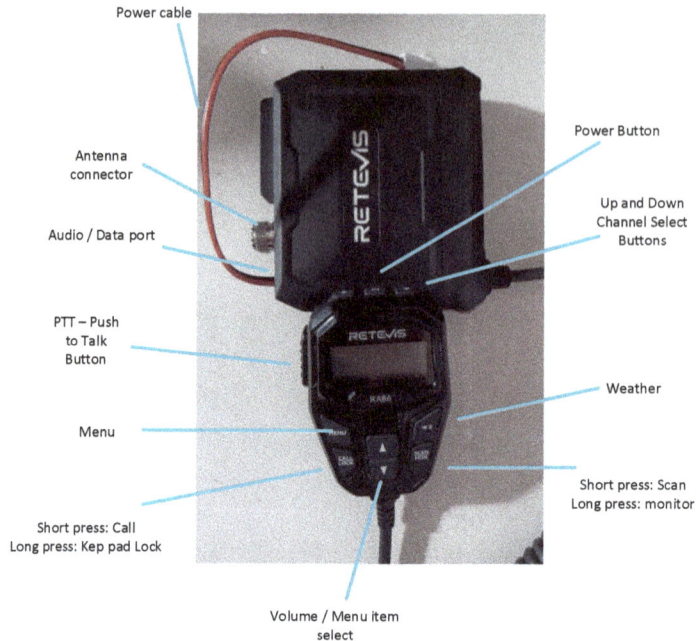

Power cable

Antenna connector

Audio / Data port

PTT – Push to Talk Button

Menu

Short press: Call
Long press: Kep pad Lock

Volume / Menu item select

Power Button

Up and Down Channel Select Buttons

Weather

Short press: Scan
Long press: monitor

The compact unit makes it easy to mount in a vehicle and all the controls in the palm of your hand. The radio is very easy to set from the microphone controls. Click on the menu to cycle through the 18 menu times such as power, bandwidth, RX/ TX tone / code squelch, LCD brightness, etc. If you miss a menu item, you have to click through the menu item again. The volume up / down button changes the menu item. The repeater channel access has to be enabled. Once enabled the repeater channels are lists as r15-r22. The repeater channels are intermixed as you cycle through the channel list. i.e. 15, r15, 16, r16, 17, r17…. Etc. Since there is no talk around button, just changing the channel to the simplex channel is one click away. On the display, the repeater channels have a triangle with arrows on two sides. Think of this symbol as talking over a mountain.

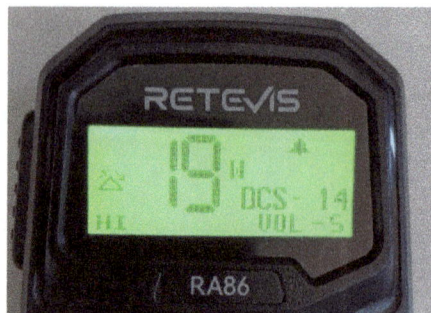

The display also shows the power output selection in the lower left. The bandwidth next to the channel number as well as a CTCSS/DCS indicator if a tone / code is set. The CTCSS/DCS is not the actual value but a code for the CTCSS/DCS tone code. The manual has all the codes for each CTCSS/DCS. Setting up the radio via the mic controller is possible, but you will need the manual to set the correct code. Finally, the channel names cannot be customized.

Click the WX button, and the radio will start scanning all 10 NOAA stations for a signal. There might be multiple NOAA stations in your area. You might have one is in English and the other is in Spanish, or you might have one station closer than another. Hit the scan button to let the radio to continue to cycle through the NOAA stations.

As far as the channel scan functions, each channel must have the scan enabled so it is part of the scan list. The scan enable for each channel can only be configured using the Mic menu controls. Channel scan enable cannot be setup in the software.

The advertised transmit power for channels 15-22 is 20 Watts. The tested output is 15 Watts. It is accepted that this is a 15-Watt radio. How the test was performed will be covered in chapter 6.

4.7.2 Retevis RA86 Programming Software
CHIRP doesn't support the RA86 as of this writing. Like the RT76P and C2, the RA86 comes with tone / code squelch preset on channels 1-22. The manual lists what each channel is set too. The software has to be used to clear out the preset tone / code squelch for each channel. The software to program the RA86 can be downloaded from the Retevis site.

1. Download the latest RA86 programming software from the Retevis website.
2. Run the installer.
3. Turn the radio on and set it to a channel that is not actively broadcasting, and then turn off the radio.
4. Connect the USB side of the programming cable to the Windows computer.
5. Open Device Manager by right click on Start button, and select Device Manager from the context menu.

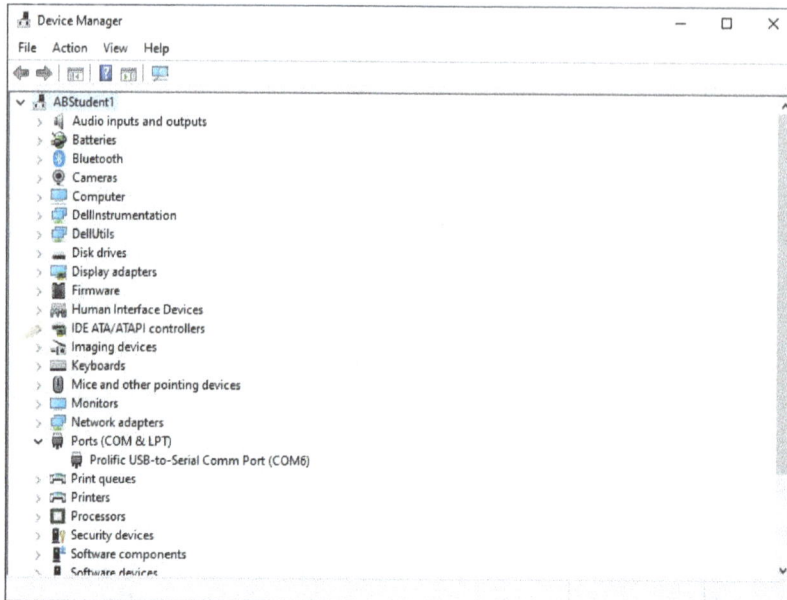

If you are using Windows 10 or Windows 11, the USB driver for the cable might already be installed. The USB cable should appear under ports. If the driver is not installed, you will have to install the driver from the CD provided with the programming cable or just run Windows update.

6. Plug the audio connector into the back data port above the power cable lead of the RA86.
7. Turn on the radio
8. Start the RA86 programming software.
9. The COM port selection is in the bottom left corner. Select the COM port that corresponds with the USB programming cable.
10. From the menu, select Program-> Read from Radio or click on the read from radio button o the toolbar.
11. Once the read finishes, click on OK button to close the dialog.

Once the data from the radio has been filled in, you will see the list of channels. Scroll down and you will see all 99 channels that can be configured. The menu on the left opens other windows to the different radio settings. The CTCSS/DCS is correctly labeled. The lower power channels 8-14 have the TX Frequency blank as these channels are only for receiving. Also you can see how the repeater channels are intermixed into the channel order, thus allow you to switch to simplex using the channel down button.

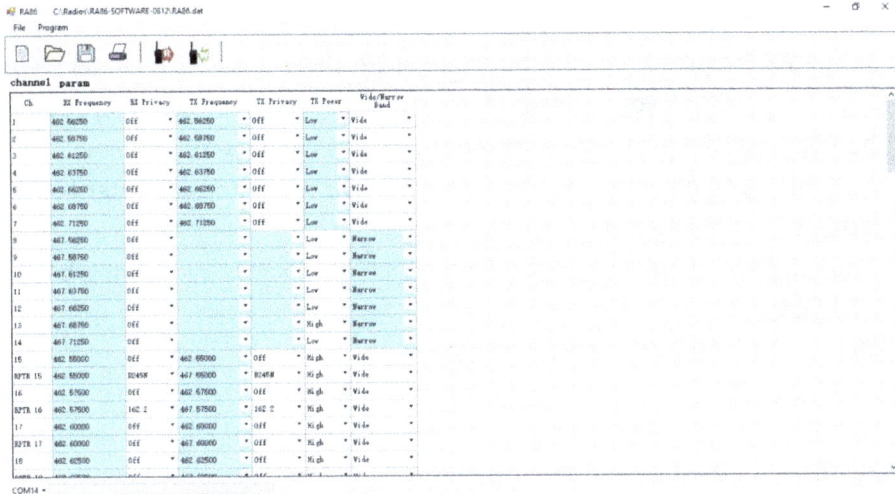

The "param" tab contains the other menu items for VOX enable, squelch settings, Mic gain, sleep, LCD brightness, etc.

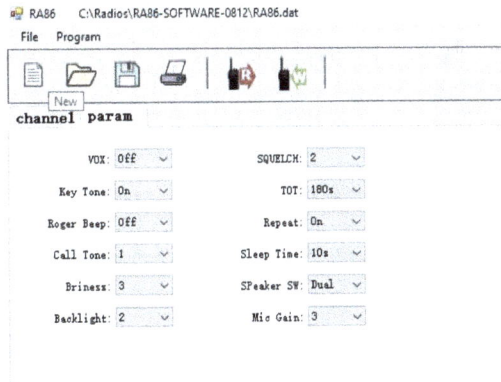

12. Once you have made all the changes, from the menu select Program -> "Write to radio…" or the write to radio button on the toolbar.
13. Click OK to close the success dialog.
14. Be sure to save the settings to a file.

4.7.3 Adding and Removing Channels

If you want to add more channels to the RA86, there is a little glitch or gotcha in the software. Let's say you want to add two more channels for different repeaters. For this example, we will set memory channels 23 and 24.

1. In the RX Frequency column for channel 23, there a GMRS repeater frequency (i.e. 462.72500). If you attempt to use the drop downs for the other settings, the software will not allow you to make any further changes for that line.
2. The trick: with the cursor still in the RX Frequency column, hit Enter key and the other settings will appear.
3. You can now use the drop down for the TX Frequency, privacy codes, bandwidth, and power.
4. Repeat these steps for other channels.

The radio knows the difference between simplex and repeater based on the 5 MHz frequency offset between RX and TX frequencies.

RA86   C:\Radios\RA86-SOFTWARE-0812\repeater test.dat

File   Program

channel param

Ch.	RX Frequency	RX Privacy	TX Frequency	TX Privacy	TX Power	Wide/Narrow Band
21	462.70000	Off	462.70000	Off	High	Wide
RPTR 21	462.70000	Off	467.70000	Off	High	Wide
22	462.72500	Off	462.72500	Off	High	Wide
RPTR 22	462.72500	Off	467.72500	Off	High	Wide
23	462.72500	103.5	467.72500	100.0	High	Wide
24	462.70000	DO23N	467.70000	DO23N	High	Wide
25						
26						

To remove a memory channel:

1. Click on the RX frequency of the channel you want to delete.
2. Delete the RX frequency
3. Hit enter. All parameters are removed.

Let's say you delete channel 23 per the picture above. Channel 24 will remain and will not move up.

4.8   Mobile Base Station Power Supply

The transmit power of mobile radios make them ideal for use as a base station. You can have a GMRS mobile radio at home that reaches family members with GMRS handheld radios. All mobile radios require a 13.8 power supply. Many HAM radio stores offer a special power supply that creates a clean power signal for mobile radios. Some radios come with a cigarette power port, which is perfect for the cigarette power plug that comes with some mobile radios.

Here is a small list of radio power supplies:

Manufacture	Models
Alinco	DM-33OMV
BTRECH	RPS-30M, RPS-30PRO
Jetstream	JTPS30LCD
Pyramid	PSU990KX
<Various>	DWC30WIN

There are specially designed cases to hold the power supply and mobile radio, and some manufactures offer a base station that combines GMRS radio and power supply into a single product:

- Hardened Systems offers an ammo can for the Midland MXT500 radio
- Retevis RMB87 combines the RA87 40W radio with desktop power supply into a single case

4.9   Radio Miscellaneous Items

The basic setting of channel, squelch, tone / code squelch, monitor, and talk around have been covered. There are other optional settings that radios can offer. This section will cover some of these features.

4.9.1   Alarm

The alarm can be set to local or send out. When set to local the alarm makes a siren noise that can be heard loudly on the speaker. This lets people nearby help locate you if you are in distress. When set to send out, the siren will be broadcasted on the current channel. Hopefully, someone is listening and can respond. Pressing the PTT kills the alarm.

There are other alarms in the radio for total talk time.

4.9.2   ANI Match

ANI is automatic number identification. You can program radios with a local ID or ANI-ID, which is a 6 digital code greater than 101. If ANI match is enabled on a channel, the radio will turn on a speaker when the ANI-ID from the send is matched on the group list.

4.9.3   Busy Channel Lockout (BCL) or Busy Lockout

Since GMRS channels are shared and tone /code squelch can be implemented, it is possible to attempt to transmit while a transmission is already in progress. This would create interference on the channel. The radios covered in this chapter have LCD screens the light up when there is a transmission coming in. Some GMRS don't have a display. The busy channel lockout function, blocks the radio from transmitting when there is another conversation going one.

4.9.4   Compand

When the option is enabled, the audio is compressed to help with the signal to noise ratio. For you to send out a compressed message means the other radios must have compand also enable to decode the audio.

4.9.5 DTMF

Dual Tone Muli Frequency (DTMF) was developed for touch tone telephones. There are the sound you hear when you press a button on your phone. There are 16 low and high frequency pairs associated with all the numbers, #, *, and phones that A, B, C, and D buttons.

Frequency (Hz)	1209	1336	1477	1633
697	1	2	3	A
770	4	5	6	B
852	7	8	9	C
941	*	0	#	D

Keypads on radios like the RT76P and C2 can send out these tones while you are pressing the PTT button. The C2 can show the decoded message on the screen.

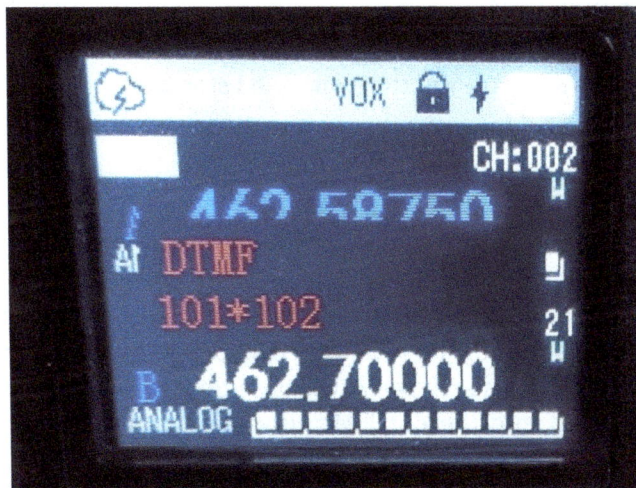

Some radios with call groups allow to setup stun or deactivate codes. Some repeaters can be remotely accessed using DTMF to enable or disable features. One of the GMRS reviewers, listed early in the chapter, developed a DTMF controller for Retevis repeaters. There are popular DTMF decoder chips like the M8870 that can be used in electronic projects, thus creating GMRS and electronic project possibilities.

4.9.6   PTT-ID /ANI
The ANI-ID can be sent as DTMF output signal before transmission, after transmission, or both.

4.9.7   Scramble
Per FCC Part 95, GMRS radios cannot scramble or encrypt message transmissions. If there is a scramble setting in the radio or software, it will probably not work.

4.9.8   Roger Beep
When enabled, a beep tone can be heard at the end of a transmission. This allows someone to know you have finished your transmission.

4.9.9   VOX
Voice activated transmit (VOX) allows for hands free operation. The option lets you set the gain of the VOX. When you speak, the radio picks up your voice and starts transmitting. The radio stops transmitting when you stop talking.

4.10 Summary
With so many features and functions to choose from, selecting the right GMRS radio can lead to analysis paralysis. The main question to be addressed is what the radio is going to be used for? You might purchase more than one radio to address a couple different use cases. There are a number of online reviewers who provide their insights on the radios they receive or purchase. These reviewers will base their review on sue and function of the radio. Diving deeper into the details is just as important. The user manuals and programming software also provides valuable information. Not everything gets documented or can be reviewed online. The chapter covered dug a little deeper into 4 radios to offer some insights and tips based on the author's testing.

5 Working with Repeaters

As the name implies a Repeater simply retransmits a received transmission. A repeater's main function:

- Extend the range of the lower powered GMRS radios
- Gets around radio transmissions blocked by terrain when placed high on hill or mountain
- Offer wireless communication in an area that lacks cell towers or has poor cell coverage

This chapter is broken into parts. The first part looks at how to connect your GMRS walkie talkie or mobile radio to a local repeater. The second part looks at repeater hardware and configuring a repeater.

5.1 Repeater Frequencies and Tone / Code Squelch

Chapter 3 covered the GMRS channels and frequencies. The repeater frequencies for walkie talkies and mobile radio are configured as follows:

Channel	TX Frequency (MHz)	RX Frequency (MHz)
23 (R15)	467.5500	462.5500
24 (R16)	467.5750	462.5750
25 (R17)	467.6000	462.6000
26 (R18)	467.6250	462.6250
27 (R19)	467.6500	462.6500
28 (R20)	467.6750	462.6750
29 (R21)	467.7000	462.7000
30 (R22)	467.7250	462.7250

The transmit and receive frequencies are split with a 5 MHz offset. The radio operates in half duplex mode when connected to a repeater. A repeater switches the transmit and receive frequencies. The frequencies in the repeater are as follows:

Channel	TX Frequency (MHz)	RX Frequency (MHz)
1	462.5500	467.5500
2	462.5750	467.5750
3	462.6000	467.6000
4	462.6250	467.6250
5	462.6500	467.6500
6	462.6750	467.6750
7	462.7000	467.7000
8	462.7250	467.7250

The repeater's receive frequency matches the transmit frequency of the walkie talkies and mobile radio, and the repeater's transmit frequency matches the receive frequency of the walkie talkies and mobile radio. Since a repeaters job is to simply be a repeater, only 8 channels are needed. Even though there are 8 channels, the repeater is always set to one channel and one channel only. You are not going to change channels like a walkie talkies or mobile radio. CTCSS/DCS code are typically added to a repeater channel.

5.2 HAM and GMRS Similarities

HAM's use repeaters on the VHF and UHF frequencies the same was GMRS use repeaters. Most people think of HAM operators sitting at a desk and communicating half way around the world. This is true when broadcasting in HF. On the higher frequences, 2 Meters (144.0 MHz to 148.0 MHz), 1.25 Meters (222.- MHz to 225.0 MHz), and 70 cm (420.0 MHz to 450.0 MHz), repeaters are needed to extend the range. VHF and UHF HAM radios come as walkie talkies or mobile radios just like GMRS.

5.3 Repeaters in Your Area

Once you get your radio, you will want to configure it to the repeaters in your area. The first step is to find GMRS repeater, and once you find one determine the rules and cost for accessing the repeater.

5.3.1 myGMRS.com

The biggest website covering GMRS repeaters is myGMRS.com. The website contains a list and map of registered GMRS repeaters across the United States and United States territories. To be listed on the website, individual repeater owners have to register. There might be more repeaters in your area that are not listed.

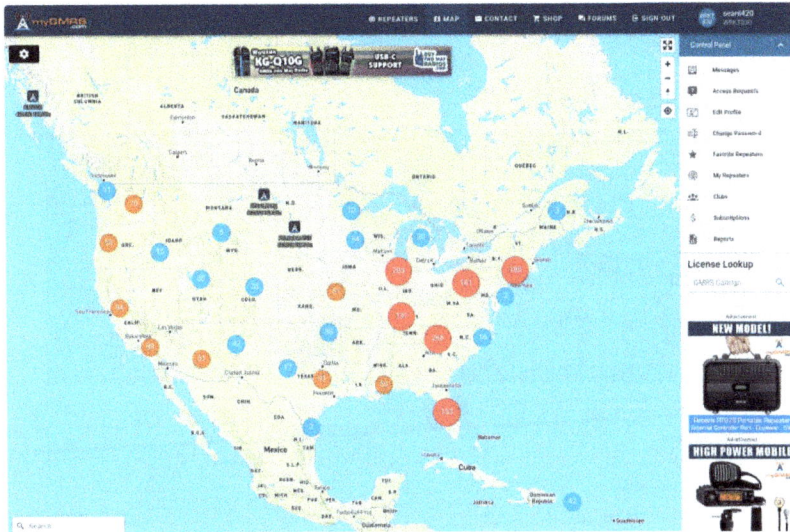

When you hover over a repeater, the map will show an estimated coverage area. Click on the repeater and you will get the connection details.

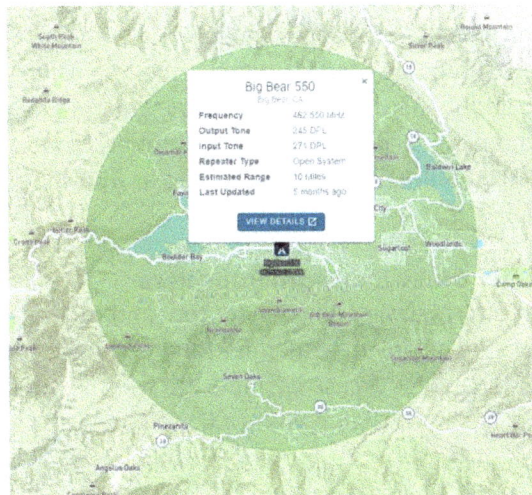

Click on the View Details button, and a new screen appears with information the repeater name, repeater owner, frequency, tone / code squelch parameters, estimated range, and rules for using the repeater.

There is some lingo used when talking about repeaters. If you are in an area with multiple repeaters, the repeaters should be on different channels so as not to interfere with one another. When talking about the different repeaters, you will hear people refer to the repeater by their last three digits of the frequency number. For example, a repeater that is on channel 27 / r19 462. 650MHz will be referred to as 650-repeater.

5.3.2   Private / Repeater Clubs and Open Repeaters

Some repeaters are open to everyone to connect too, and others are for members only. The repeaters that are listed as private / member only will have the tone / cope squelch unlisted. You will have to request access or pay a fee to access the repeater. Only then will you get the tone / codes to configure your radio to connect to the repeater.

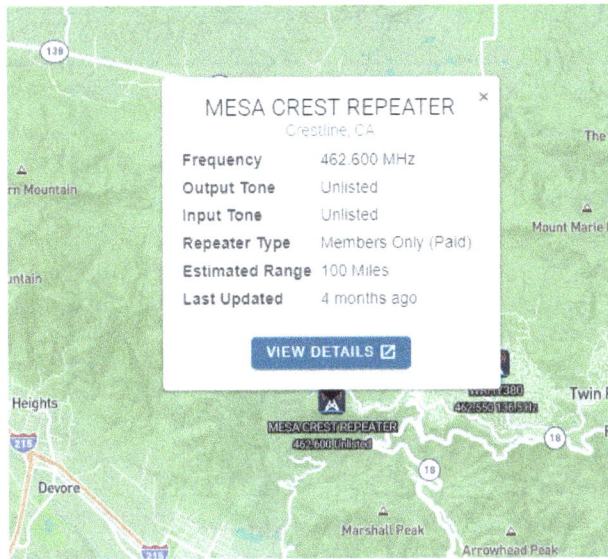

Open repeaters will state they are open and list the tone / code squelch information. There are rules about accessing the repeater. Regardless if the repeater is private or public, it is best to request access to get further information for the repeater owner or just to check to see if the repeater is still operational.

5.3.3   Other Sites and Linking Repeaters

Since HAM operators can access repeaters, there are other websites that list repeaters for HAM radios. Repeaterbook.com is major website for amateur radio, which includes some GMRS repeaters.

GMRS repeaters CANNOT be linked to one another. It would defeat the intended purpose for local communication if GMRS repeaters were connected to other repeaters hundred to thousand miles away. If you want to use a radio for long distance communication, you should get a HAM license.

5.3.4   Example Connecting to a Repeater

Let's walk through a couple examples of connecting to a repeater. The key is understanding the difference of input tone and output tone as listed on the website. Input tone is the repeaters receive CTCSS/DCS tone / code. Output tone is the repeaters transmit CTCSS/DCS tone / code. For your radio, you set the TX CTCSS/DCS in your radio to the input tone for the repeater. Set the RX CTCSS/DCS in your radio to the output tone of the repeater.

First example: repeater with the same tone / code squelch information for RX and TX. Frequency is 462.650 Mhz. The Input Tone is 072 DPL and the output tone is 072 DPL. DPL term is a bit confusing as it is an old term – digital private line. The 072 is a DCS code. CTCSS tones would have a tone frequency and Hz in the listing. In your radio menu or via the software, configure channel 27 / r19 with RX DCS set to D072N and TX DCS set to D072N. Typically, GMRS uses the N and not the I for the DCS. Once configured, press and release the PTT without say anything into the radio, the repeater my send a message back in the form of a tone, morse code or voice. Press the PTT, speak your call sign, say that you are testing the repeater connection. Hopefully, someone is on the repeater and can respond.

Second example: repeater with the different tone / code squelch information for RX and TX. Frequency is 462.6575 Mhz. The Input Tone is 155 DPL and the output tone is 245 DPL. The input tone is the most important as it is need to transmit to the repeater. In your radio menu or via the software, configure channel 24 / r16 with RX DCS set to D245N and TX DCS set to D155N. Once configured, press and release the PTT without say anything into the radio, the repeater my send a message back in the form of a tone, morse code or voice. Press the PTT, speak your call sign, say that you are testing the repeater connection. Hopefully, someone is on the repeater and can respond.

5.3.5   Road Trips
The last chapter pointed out that the different radios have extra memory channels available. If you are planning a road trip, you can use myGMRS.com to pre-program the extra memory slots with the repeaters along your route.

5.4   Repeater Architecture and Manufacturers
A repeater's architecture and setup are designed to split the TX / RX frequencies. The GMRS radio transmits and receives on separate frequencies. A duplexer is put in place so only one antenna is needed for the both TX and RX. A duplexer attenuates the TX transmission from spilling over to the RX receiver and damaging the receiver.

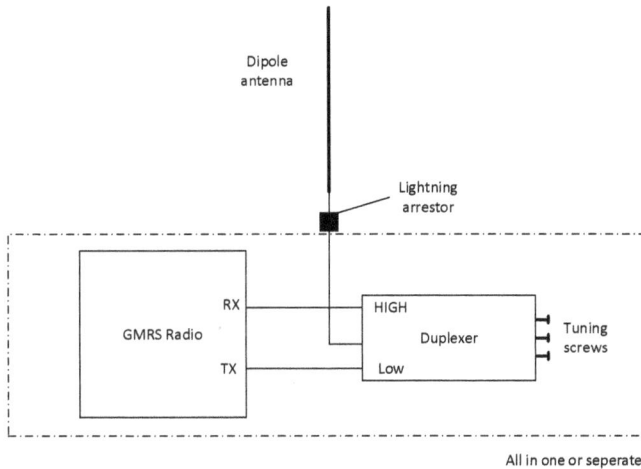

Dipole antenna

Lightning arrestor

GMRS Radio

RX

TX

HIGH

Duplexer

Low

Tuning screws

All in one or seperate

If you are interested in setting up your own repeater, you won't break the bank as repeaters come in different shapes and sizes. Some repeaters are as small as a suite case, and others are mounted in a server rack. There are repeater solutions that connect two mobile radios together to create a repeater. Here is a list of known repeater manufactures at the time of this writing.

Manufacture	Website	Model
Bridgecom Systems	bridgecomsystems.com	BDR-4500 UHF
BTECH	baofengtech.com	GMRS-RPT50
BuyTwoWayRadios	buytwowayradios.com	2 KG-1000G mobile radios
Hytera	hytera.us	HR1062H
Icom	icomamerica.com	FR5300, FR6300
Midland	midlandusa.com	MXR10
Motorola	motorolasolutions.com	MOTOTRBO SLR 5700
Retevis	retevis.com	RT97L, RT92S, RAR87 (2 RA87 mobile radios), R1
Wouxun	wouxun.com	KG-DR3000, KG-DR2000, KG-DR1000 (2 mobile Radios),

The Midland MXR10 and the Retevis RL97L have the radio and duplexer together in a single package. These repeaters are only 25W transmit power. Some repeaters have duplexers built-in, some don't. Here is a list of GMRS duplexers for 50W transmit power:

Manufacture	Website	Comments
Bridgecom Systems	bridgecomsystems.com	BCD-440 UHF Mobile-Type Duplexer
FUMI		SGQ-450A
XLT Communications (BuyTwoWayRadios)	buytwowayradios.com	DP-GMRS 50

Tuning the duplexer is the work that has to be performed. Some retailers will tune the duplexer for you.

Warning: If you do install a repeater in your house, radio shack, man cave, she shed, etc. and the antenna outside, you should install a lightning arrestor in the antenna feed line coming into the building. The lightning arrestor will prevent damage to equipment and property.

5.5   Two Mobile Radios as a Repeater

Some of the higher end 50W GMRS repeaters can cost some money. There is an alternative solution by connecting two mobile radios together. The Wouxun KG-1000G and the Retevis RA87 have a repeater capable feature. BuyTwoWayRadios has a nice video demonstrating how to connect two Wouxun KG-1000Gs together into a radio. Retevis sells the two RA87's and the power supply into a single product called the RAR87. The RA87 manual walks through the simple steps to configure the two RA87s as a repeater.

In both cases, a cable is connected between the radios to establish the repeater function. One radio is set to the repeater low TX frequency and the other is set for the higher RX frequency. Without a duplexer, two antennas are needed, and the antennas need to be placed far apart so the TX doesn't interfere with the RX.

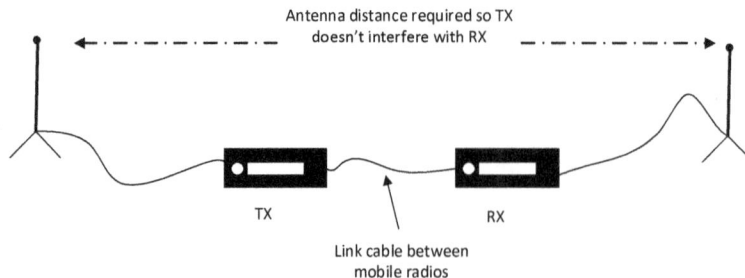

Antenna distance required so TX doesn't interfere with RX

TX   RX

Link cable between mobile radios

A duplexer is a better solution so only one antenna is required. Two mobile radio repeater gives you a high power (40W to 50W) repeater solution that doesn't break the bank.

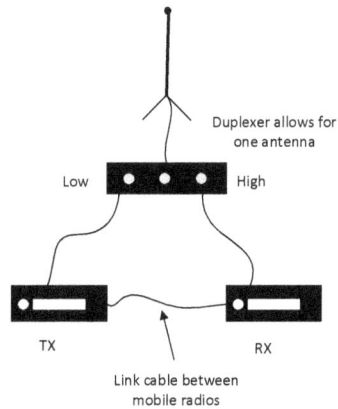

Note: A Wouxun KG-1000G and a Retevis RA87 are different products from different manufactures. They cannot be connected together as a repeater since their data links are incompatible.

5.6 Retevis RT97L

The Retevis RT97L is a 25W portable GMRS repeater. The RT97L is also sold as a UHF repeater. Retevis also has the RT97S, which is an older version. RT97L's portable design is suitable for different use cases:

- Family Farm
- Emergency response center
- Base camp

5.6.1 Features and Tips

The RT97L has the following features:

- 25W high power output
- Upgraded internal duplexer
- Portable and can be wall mounted
- Can be powered in a vehicle with a cigarette lighter plug
- IP66 water proof
- High and low temperature protection

- Can connect to a Raspberry Pi via the DB9 port.
- Some basic configuration can be performed via the onboard menu / LCD screens
- Software program required to set frequencies and CTCSS/DCS codes. USB to DB9 programming cable is included.

Retevis also sells an antenna, cable, and a microphone for the RT97L. The microphone connects to the DB9 port, and can be used to transmit out on the repeater channel.

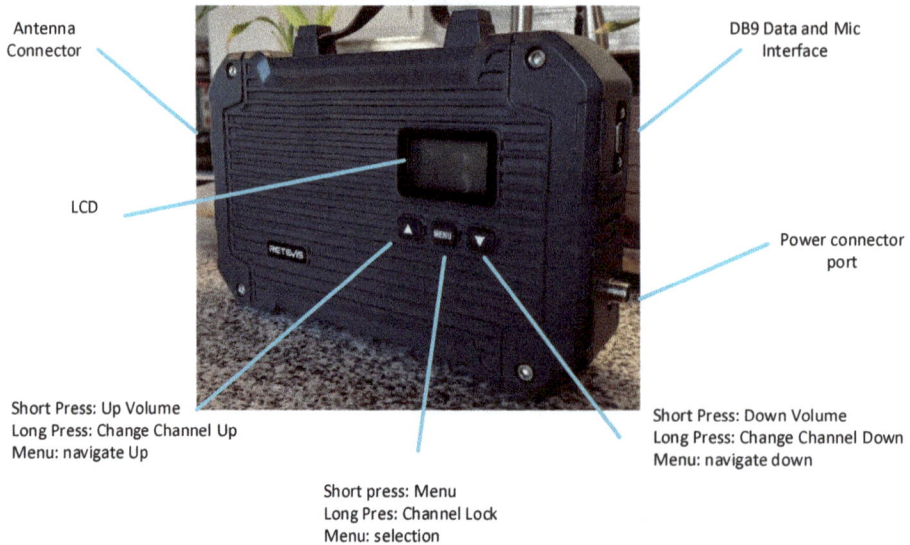

Antenna Connector

DB9 Data and Mic Interface

LCD

Power connector port

Short Press: Up Volume
Long Press: Change Channel Up
Menu: navigate Up

Short Press: Down Volume
Long Press: Change Channel Down
Menu: navigate down

Short press: Menu
Long Pres: Channel Lock
Menu: selection

The even though the repeater has 16 channels, only one channel can set as the repeater channel. The Relay function needs to be set to "ON" for the repeater to operate as a repeater, and whatever channel is show on the LCD will be the repeater channel.

Note: There is a firmware upgrade for the RT97L. The documentation that comes with the firmware install covers the steps well enough.

5.6.2   Retevis RT97L Programming Software

The only way to set the CTCSS/DCS codes is using the programming software. All channels are set to CTCSS 136.5 Hz.

1.   Download the latest RT97L programming software from the Retevis website.
2.   Run the installer.
3.   Connect the USB side of the programming cable to the Windows computer.
4.   Open Device Manager by right click on Start button, and select Device Manager from the context menu.

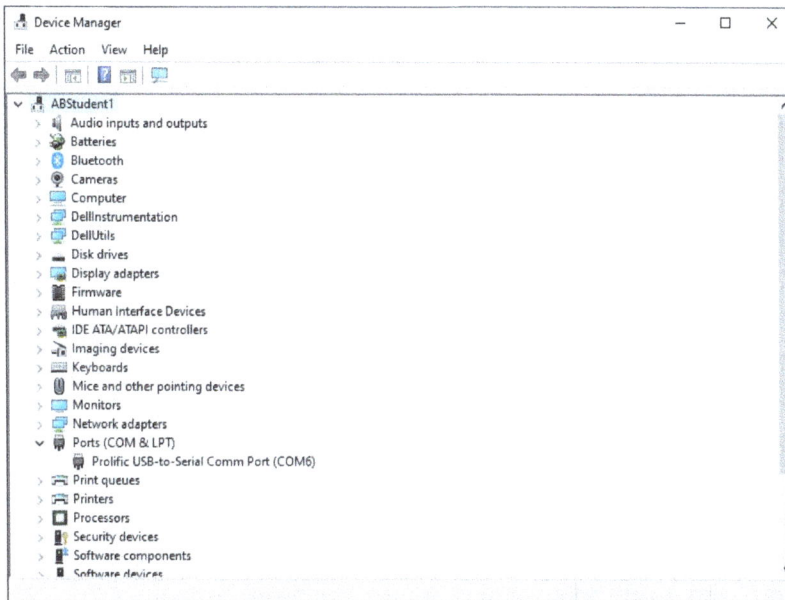

If you are using Windows 10 or Windows 11, the USB driver for the cable might already be installed. The USB cable should appear under ports. If the driver is not installed, you will have to install the driver from the CD provided with the programming cable or just run Windows update.

5.   With the RT97L unplugged / unpowered, plug the DB9 connector on the cable to the DB9 port on the RT97L.
6.   Power on the RT97L.
7.   Start the RT97L programming software.

8. Click on the COM port button in the tool par to set the COM port of the USB cable.
9. From the menu, select Program-> Read from Radio or click on the Read radio button on the toolbar.
10. Once the read finishes, click on OK button to close the dialog.

Once the data from the radio has been filled in, you will see the list of channels. There are 16 channels broken into 8 repeater channels set to wideband and 8 repeater channels set to narrowband.

--- RT97L-C:\Radios\Retevis RT97L GMRS Repeater\basic.txt — □ ×

File   Machine   Edit   Program   Set   About

NEW   OPEN   SAVE   READ   COM   WRITE   About

CH	Rx Freq	Tx Freq	CT/DCS Dec	CT/DCS Enc	Tx Power	W/N	SCAN ADD
1	467.55000	462.55000	136.5	136.5	High	Wide	Del
2	467.57500	462.57500	136.5	136.5	High	Wide	Del
3	467.60000	462.60000	136.5	136.5	High	Wide	Del
4	467.62500	462.62500	136.5	136.5	High	Wide	Del
5	467.65000	462.65000	136.5	136.5	High	Wide	Del
6	467.67500	462.67500	136.5	136.5	High	Wide	Del
7	467.70000	462.70000	136.5	136.5	High	Wide	Del
8	467.72500	462.72500	D131N	D131N	High	Wide	Del
9	467.55000	462.55000	136.5	136.5	High	Narrow	Del
10	467.57500	462.57500	136.5	136.5	High	Narrow	Del
11	467.60000	462.60000	136.5	136.5	High	Narrow	Del
12	467.62500	462.62500	136.5	136.5	High	Narrow	Del
13	467.65000	462.65000	136.5	136.5	Low	Narrow	Del
14	467.67500	462.67500	136.5	136.5	High	Narrow	Del
15	467.70000	462.70000	136.5	136.5	High	Narrow	Del
16	467.72500	462.72500	136.5	136.5	High	Narrow	Del

Version0.1

The optional features contain items for squelch, mic volume, starting channel on boot, Relay on/ off, and relay delay timing. Delay timing is something to test to see if communication is clearer and the system is not going to overheat.

```
Optional Features                                                    ✕

        Squelch 2        ▾    Relay Delay 1           ▾
         Volume 8        ▾          Relay ON          ▾
   Audio output 5        ▾       Language English     ▾
       Mic Gain 0        ▾            STE frequency   ▾
        Scan CH OFF      ▾   Channel Mode CH+Frequency ▾
        Channel 8        ▾
   Channel Lock Manual   ▾
     Back light AUTO     ▾
Low temperature OFF      ▾


            ┌──────────┐              ┌──────────┐
            │    OK    │              │  Cancel  │
            └──────────┘              └──────────┘
```

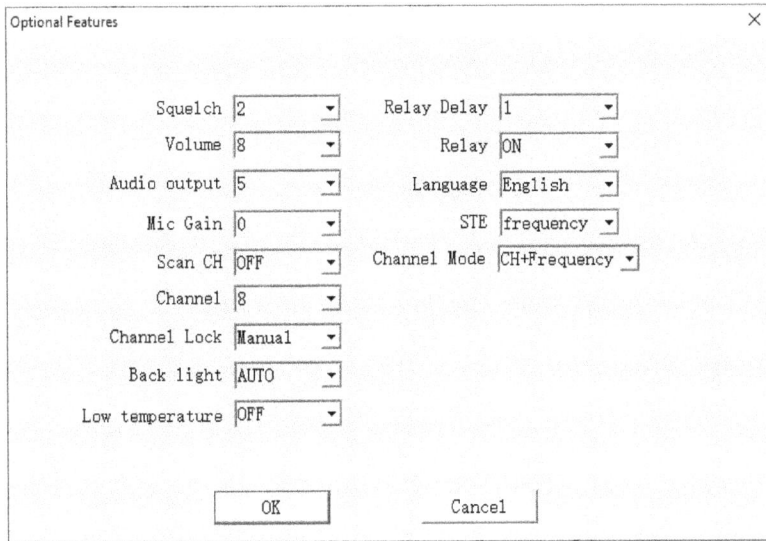

11. Once you have made all the changes, from the menu select Program -> "Write to radio..." or the Write button on the toolbar.
12. Click OK to close the success dialog.
13. Be sure to save the settings to a file.

5.6.3   Example of Tone/Code Squelch (Not Privacy Codes) in Action Part 2

Chapter 3 had an example showing the tone / code squelch was not private. For this example, radio A and B are connected to the repeater using the DCS code D131N on Frequency 467.725. Radio C is channel 22 and has no DCS codes configured.

- Radio A and B can communicate with each other through the repeater.
- Radio C can hear the communication between A and B since the RX tones/codes are turned off.
- When Radio C transmits, neither Radio A or B can hear Radio C's transmission.

Repeater
RX Frequency: 467.7250 CTS/DCS: D131N
TX Frequency: 462.7250 CTS/DCS: D131N

Radio A
Channel 30/r22
RX: CTS/DCS: D131N
TX: CTS/DCS: D131N

Radio B
Channel 30/r22
RX: CTS/DCS: D131N
TX: CTS/DCS: D131N

Radio C
Channel 22
RX: CTS/DCS: OFF
TX: CTS/DCS: OFF

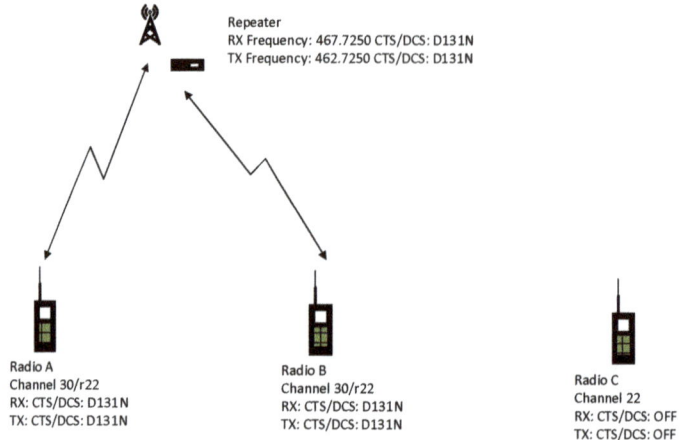

Like the example in chapter 3, CTCSS / DCS Tone / Codes are not "private".

5.6.4   Example of Talk Around

Building on the last example, let's say Radio A is the Retevis C2 and Radio B is the Retevis RA86 mobile radio. The RA86 doesn't have a talk around functions, but the CS does. If both radios are close enough to where they can talk clearly without the repeater, the C2 can be set into talk around mode and the RA86 can just change to channel 22 with one click on the mic controls. Both radios can talk without the need for the repeater.

5.6.5   DB9 Port Projects

The advertised features call out connecting a Raspberry Pi to the DB9 of the RT97L. Retevis has released the connection signals of the DB9 port.

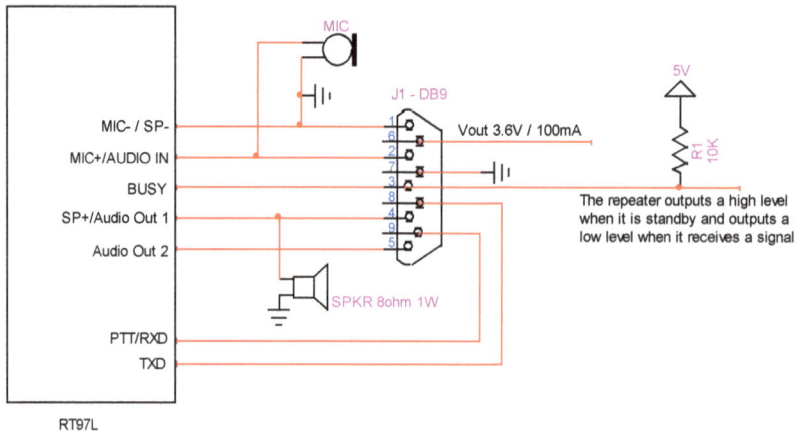

MIC

J1 - DB9

MIC- / SP-

MIC+/AUDIO IN   Vout 3.6V / 100mA

BUSY

SP+/Audio Out 1

Audio Out 2

5V

R1 10K

The repeater outputs a high level
when it is standby and outputs a
low level when it receives a signal

SPKR 8ohm 1W

PTT/RXD

TXD

RT97L

The MIC and speaker shown in the circuit is how the hand-held mic is connected. The DB9 port provides the opportunity to create add-ons that work with the repeater:

- Repeater ID - One company developed a solution to broadcast a repeater ID. You can find the product from Repeater ID at repeaterid.com.
- Interface board and software from GMRS _two_way_radio: www.gmrstwowayradio.com/store/shop/
- Raspberry Pi – GitHub project for the Raspberry Pi: github.com/emuehlstein/pyrepeater, although this appears to support the RT97S.
- Audio recording device to monitor communication over the repeater
- Created a DTMF decoder board so you can send remote DTMF signals and have the custom board perform different actions like report the ambient temperature at the repeater.

5.7 Repeater Setup Etiquette

If you do setup your one repeater, to prevent interference with other repeaters be sure to use a repeater frequency that is not used by another repeater in your area. In the case of the RT97L, you can switch the repeater frequency to low power to reduce the ranged covered.

5.8 Summary

GMRS Repeaters have a simple but powerful function. The ability to extend the communication range of GMRS walkie talkies and mobile radios from a few miles to 20-50+ miles make GMRS an effective communication solution for a local area that might not have cellular coverage. The GMRS community is growing and so are the number of repeaters. myGMRS.com is the go-to site to find repeaters in your area. Once you have connected an new world opens up.

6 Radio Tools and Diagnostics

GMRS as a gateway into Amateur (HAM) Radio is the book's driving theme. HAM radio and GMRS are similar in a couple ways. The first is that GMRS radio can connect to a repeater just HAM radios 2M/70cm bands. The second similarity is that the antenna can be replaced. Unlike FRS radios, GMRS radios allow you to change out the antenna. HAM operators are very familiar with changing and working with different antenna types. There are several books and on radio antennas and computer programs to model to build custom antennas. The American Radio Relay League (ARRL) has a complete book on antennas. GMRS radios only have one type of antenna, but the characteristics of an antenna can change the radios performance. Choosing the wrong antenna or improperly tuned antenna can damage the radio. GMRS radios do have circuity in them to prevent issues, but anything can happen. The same tools that HAM operators use to tune their radios are the same tools that can be used with GMRS radios. This chapter will first look at antenna basics and then dive into the tools to make sure the antenna is properly tuned to the radio.

6.1 Antenna Basics

Antennas are transducers. Like a speaker that converts electrical signals into audible sound waves, antenna take the electrical signal and converts them into EM waves. The sinusoidal wave going into the antenna makes the electrical charge oscillates in the antenna. The EM wave radiates out away from the antenna.

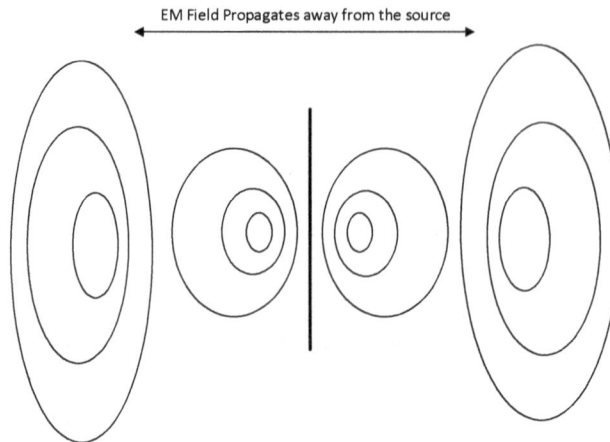

EM Field Propagates away from the source

Antennas come in different types such as loops, dishes, and Yaggi. The basic radio antenna is a ½ wave dipole, where the feed line comes into the middle of the antenna.

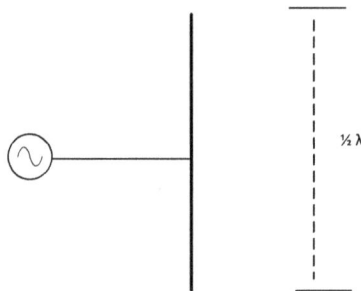

½ λ

GMRS antennas are ¼ wave vertical mono pole. The radiation pattern from the radio is omnidirectional. The length of the antenna is can be found using the following equation:

$$\text{Antenna Length (meter)} = \lambda/4 = c / (\text{freq} * 4)$$

$$\text{For GMRS, Antenna Length} = 300 / (462*4) = .162 \text{ meters or 6.5 inches.}$$

This is why we see small rubber ducky antenna come with walkie talkies. The longer the antenna the better, thus antenna manufactures will add more windings inside the antenna to get length and improve the gain. Since a dipole is the fundamental radio antenna, the ground plane makes up the other half of the antenna.

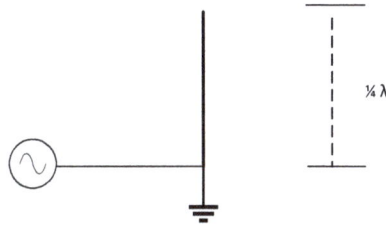

The antenna is not just used to transmit, but it is also used to receive signals as well. If you ever played with a crystal radio, the antenna coil and variable capacitor in parallel make up a tank circuit. Adjusting the variable capacitor tunes into the AM broadcast frequency.

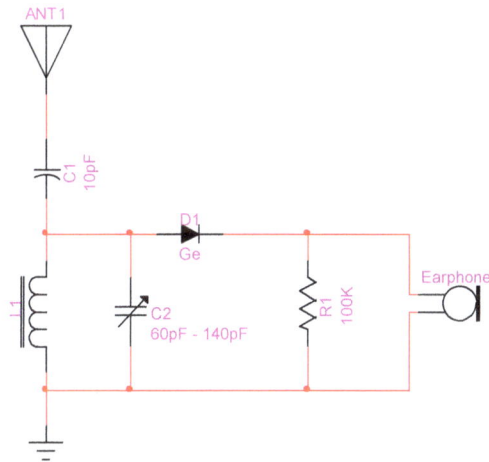

The germanium diode rectifies the signal, which recovers the audio modulated signal. You will notice that there is no battery in the crystal radio as all the power is coming from the EM wave. The inductor / capacitor combo (tank circuit) leads us to the critical importance of tunning the antenna. The component characteristics of an antenna are a combination of resistor, capacitor, and inductor.

6.2   50 Ohms Impedance and Matching

We will save you from complex mathematics and 4-year engineering degree to get to the point. Electrical Engineering courses first cover the basic Kirchhoff's Current and Voltage Laws for DC circuits. When it comes to sinusoidal signals, circuit analysis using KVL or KCL is very involved. Since sinusoidal signals are the same frequency throughout the circuit, the magnitude and phase angle are changing. Since complex numbers have the same characterizes of a sinusoidal signal with a magnitude and phase angle, complex numbers are used to analyze time vary circuits. Resistors follow Ohms law regardless of DC or sinusoidal signals. Inductors and capacitors have a characteristic impedance that varies over frequency. The mathematical representation of the impedance is defined as a complex number with a real part and imaginary part.

The nature of inductors and capacitors are to store energy. When a current is applied, the inductor builds up a magnetic field while the capacitor will be up a charge. When the current is removed, both components will return the energy back. The current will go to the path of least resistance. In a radio, we want 100% of the signal to go out the antenna. Of course, nothing is ideal. Between the transmitter circuitry and the antenna, there are connectors, feed wires, and imperfections in materials that add impedance. Any mismatch in impendency between the transmitter circuitry and the antenna will have energy return to the transmitter, which is okay in small amounts, but a larger return can do damage to the radio.

All antennas and coaxial cables for radios and TV specify 50Ω impendence. 50Ω is not a random choice. Coaxial cable was first used in the 1860s for undersea cable transmission. Bell Laboratories in 1929 experimented with different impedances for coax cable. They found that 77Ω was ideal for low attenuation and 30Ω was perfect to deliver high power. If we take the middle value between the two and some rounding, 50Ω provide to be the best balance between attenuation and power delivery.

With all that said, the important radio antenna test is check for impedance mismatch. Special test equipment checks the reflected wave back to the transmitter in the form of the standing wave ratio (SWR or VSWR). Ideally, the SWR ratio should be 1:1, but nothing is perfect so SWR ≤ 1.50 is acceptable. We will look at a couple of these test tools later in the chapter.

6.3 Speaking in Decibels

Decibels are logarithmic scale to measure intensity either gain (+dB) or loss (-dB). 0dB is the base line for what a human can hear. You will see on sports programs or stadiums measuring the dB of the crowd noise compared to other noise sources like jet engines.

The ratio concept originated the forementioned undersee cables as a way to measure signal loss over distance. For radio, decibels are use in a few areas:

- Filter bandwidth – measure to where the signal gets attenuated or half the input power.
- Feed line loss - usually presented as dB per 100 ft or 1000 meters.
- Microphone gain in dB
- Amplifiers are measured in dB
- Transmit Power output is measured in dBm
- Antenna gain is in dB or dBi. The gain is a how well the antenna can send or receive signals in a specific direction compared to a theoretical isotropic antenna.

6.4 Connectors

The radios and repeaters discussed in the last two chapters have antenna connections of different types. Once you start getting into radios and test tools, you will get familiar with all the connector type and adaptors that are possible. Here are the common connectors:

Type-N

PL259 (PLug Male) SO239 (SOcket Female)

SMA-M and SMA-F (SMA = Subminiature) Also known as SMAP (Plug/male) and SMAJ (Jack/female)

BNC

Choosing the right connector is important as different radios can come with different connectors. Walkie talkies from the same manufacturer can feature a male or female connector.

6.5   Antenna Selection and Installation

The first few sections have covered what an antenna does, some characteristics, and the importance of impedance matching. Now, we need to look how to select and install an antenna.

6.5.1   Retevis RT76P Example

Let's start with an example. One of the first GMRS radios purchased was the Retevis RT76P. The RT76P support FM radio stations, NOAA stations, 2M VHF RX only, and all the GMRS channels. The radio couldn't pick up the NOAA stations and was only able to pick up a few FM stations. GMRS communication was fine. Since the antenna could be replaced, the logical solution was to find new antenna that could improve the receive sensitivity of the radio. The radio required a SMA-M antenna sine VHF was support the antenna had to support VHF and UHF. After a search, two replacements were found, and both solved the FM and NOAA reception issues.

6.5.2   Specifications

You may never have to replace the stock antenna with your GMRS walkie talkie, but having options can improve radio communications. There are certain specifications to look for.

- Frequency Range: GMRS UHF frequencies have to be supported. Antenna lists the range like those shown in the pictures below, or show a general UHF frequency to cover the whole range. For GMRS, this could be 430MHz

- Connector: Check to make sure the connector is correct.

- Max Power: 10W or 20W is more than enough for walkie talkies. How power rating is needed for mobile radios and repeaters.

- VSWR: Hopefully, rated to be less than 1.5.

- Gain: typically, 2.15dBi for walkie talkies

- Size: The size and the length of the antenna can matter for practical portability. The picture below shows to antennas that have similar specifications for frequency, gain, and VSWR. The only difference is power (20W versus 10W), but for a 5W walkie talkie, power doesn't matter. Both provide the same reception and communication clarity.

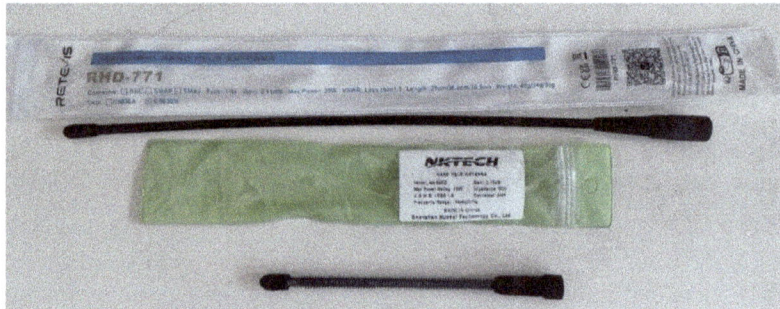

6.5.3   Installation

A walkie talkie antenna is a simple installation. Mobile radios and repeaters things can get a little more complicated. For mobile radios mounted in a vehicle, there are some general rules to follow:

- Route cables away from vehicle power cables.
- Keep cables away from other electronics like ignition starter.
- Don't wind up a cable. If the cable is too long, splice the cable to shorten and use crimp connectors to add a new connector.
- Ground the antenna. The vehicle metal acts as the other half of the antenna dipole.

- The placement of the antenna forms the EM radiation path. Placing on the top of the vehicle will have a different characteristic than placed on the hood or left/right bumper.
- If there is interference, try using ferrite cord beads.
- Check SWR

Here are some general guidelines for repeaters:

- The antenna should be in a high place to reach the widest area.
- Keep antenna away from power lines.
- The coaxial feed line should be a heavy quality like RG213.
- If the feedline is going underground, putting the feedline through a PVC pipe can deter gophers, rats, or mynocks from chewing on the cables.
- A lightning arrestor should be placed where the feed line comes into the building housing the repeater.
- Water tight connectors like Type-N or PL259/SO239 should be used for all connections.

6.6 SWR Meter

Most radios have built in auto tunning circuitry to adjust for an impedance mismatch so you don't have to worry about matching impedances. If reception is distorted or the radio is getting really hot, performing a SWR test on the antenna is a good check. The SWR readings vary depending on the frequency, which is a result of LC circuits.

An SWR meter is use to read the SWR from the antenna and forward power from the radio. The idea is that the SWR meter sits in the feed line between the transmitter receiver and the antenna. The signal coming from the radio to generate the readings.

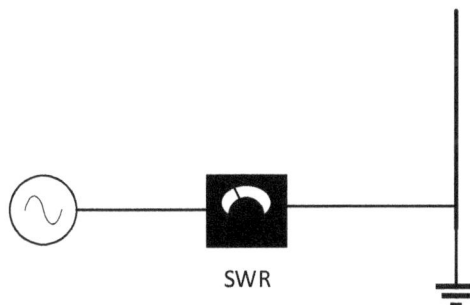

SWR

There are different SWR meters on the market, but the Surecom SWR meters get the most attention. The smaller SW-33Plus is intended for handheld walkie talkies, and the SW-102 is ideal for mobile radios and repeaters.

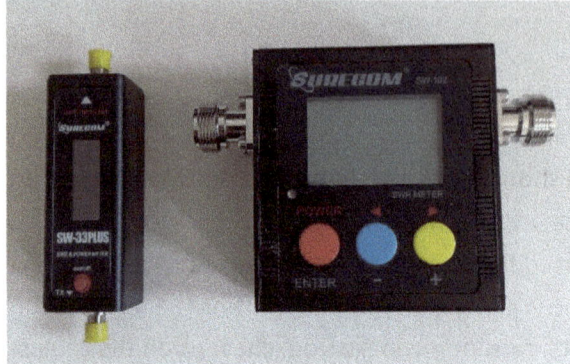

SW-33Plus comes with SMA adapter connectors, a 5W dummy load, and a USB cable to charge the internal battery. The 5W dummy load is used for measuring power output. The SW-102 has Type-N connectors on either end, and the SW-102S has SO239 connectors. Both models come with adapter connectors to change from Type-N to SO239, charging cable, and power adapter. The SW-102S is better for GMRS mobile radios and repeaters that use PL239/SO239 connectors. A dummy load for testing power is not included, but there are 50W dummy load is sold separately.

6.6.1   Using the SW-33Plus for Handheld Walkie Talkies

Testing for SWR with a walking talkie is a bit of a challenge. The radio body and physical touching of the radio make up the ground plan portion of the antenna. With the SWR meter in place, the ground plane is disconnected from the antenna.

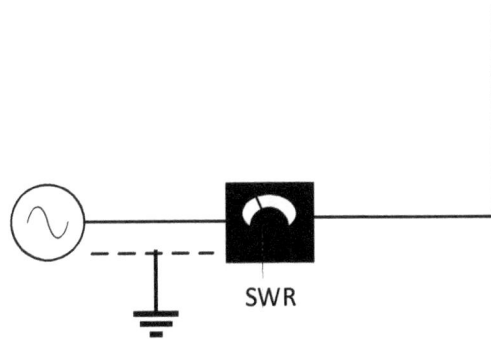

SWR

The instructions for the SW-33Plus state that you should hold the SW-33Plus with your hand to create the ground. The first test to perform is a forward power test and then test VSWR.

1. Remove the antenna
2. Connect the SW-33Plus' TX side, either directly or using the provided adapters, to the radio antenna connector.
3. Connect the dummy load to the antenna side of the SW-33Plus
4. Press and hold the red button on the SW-33 until it turns on.
5. You can tap on the red button to cycle through the different display configurations. Tap on the red button until you see "FW W" on one of the display configurations.
6. Turn on the radio.
7. Set the radio channel to 28 or 20.
8. Tap the PTT button on the radio. The SWR meter should show the forward power from the radio.

9. Now replace the dummy load with the antenna.
10. Tap the red button on the SW-33Plus until you see VSWR in one of the display configurations.
11. Hold your hand around the SW-33Plus, hold the radio vertically, and tap the PTT button on the radio to get the VSWR reading.

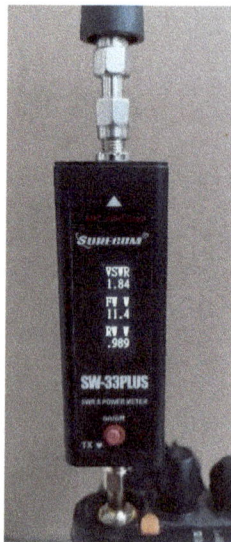

12. Let's see what happens if your hand is not holding the SW-33. Remove your hand, and tap the PTT button on the radio.

The SWR value jumps up considerably without a solid ground connection. Ground is very much part of the GMRS antenna system. The following tables lists our results for the radios discussed in chapter 4. The radios are 5W at full power. Channel 12 is one of the 7 channels that must output only 0.5 W power.

Wouxun KG-805G

Channel	Forward Power (Watts)	SWR (Stock Antenna)	No Gain antenna tested
3	5.90	1.68	
12	0.71	1.01	
19	5.90	1.58	
28	5.86	1.64	

Retevis RT76P

Channel	Forward Power (Watts)	SWR (Stock Antenna)	SWR (Retevis whip antenna SMA-M)
3	5.45	1.85	4.46
12	0.25	1.07	4.54
19	5.46	1.66	4.36
28	5.43	1.92	4.44

Retevis C2

Channel	Forward Power (Watts)	SWR (Stock Antenna)	SWR (Retevis whip antenna SMA-F)
3	5.22	1.71	2.59
12	0.42	1.02	1.04
19	5.22	1.78	2.69
28	5.25	1.83	2.91

Notice that the SWR values change over frequency. This is the LC characteristics of the antenna. The vector network analyzer tool to be discussed next will show this curve graphically.

6.6.2   Using the SW-102 for Mobile Radios and Repeaters

The SW-102 is ideal for testing mobile radios and repeaters. Depending on the version of SW-102, you will need adapters to connect the different connector types. The SW-102S with SO239 connector is a better choice since all the mobile radios and feed lines use PL259 and SO239. The SW-102 with Type-N connectors will require some extra adapters.

1. Remove the antenna from the mobile radio.
2. Connect the SW-102' TX side, either directly or using the provided adapters, to the radio antenna connector.
3. Connect the dummy load to the antenna side of the SW-102.
4. Press and hold the red button on the SW-102 until it turns on. There are some optional settings, but the defaults will be used for this test.
5. Turn on the radio.
6. Set the radio channel to 3.
7. Tap the PTT button on the radio. The SWR meter should show the forward power from the radio.

13. Now replace the dummy load with the antenna.
14. Tap the PTT button on the radio to get the SWR reading.

The following tables lists our results for the Retevis RA86 mobile radio and RT97L repeater.

Retevis RA86

Channel	Forward Power (Watts)	SWR (Stock Antenna)	SWR (Retevis M200 Gain antenna)
3	5.57	1.92	2.59
19	14.47	2.15	2.69
28	14.70	2.04	2.91

Retevis RT97L

Channel	Forward Power (Watts)	SWR (Stock Antenna)
8 on repeater - R22/30 462.7250 MHz	26.54	1.50

The Retevis MA02 FRP antenna 144/430MHz was used for the test. The ground plan is 3 rods that angle down at the bottom of the antenna.

For RT97L MA02 FRP antenna, the SWR result above is a direct connection between the feed line and the antenna. If a connector for a mobile stand is used, the SWR can go up to 1.70, which is still acceptable. The manufacture's recommendation for the MA02 FRP is to clamp the antenna to a pole and connect the feedline directly to the antenna. A lightning arrestor was not added to the test.

We see that the SWR on the antennas are not a perfect 1.00. Connectors and feedlines to perform the test can add to the readings. Again, built in antenna tuners can compensate. SWR changes over frequency is clearly shown in the results, which brings is to the next test tool.

6.7   Vector Network Analyzer (VNA)

The Surecom SWR / Power meters rely on the radio to produce the signal. Another test tool called the Vector Network Analyzer (VNA) is used to test radio circuits such antennas, filters, attenuators, amplifiers, etc. Where the SWR meter tests individual frequencies one at a time, the VNA tests a range of frequencies in one sweep. The VNA provides the signal to test the circuit, thus isolating the device under test (DUT) from the radio. Testing the antenna SWR with VNA only requires the antenna.

6.7.1 nanoVNA

VNA prices can range from $2,000 to $75,000+, which is cost prohibitive for any hobbyist. Technology has improved to allow cheaper alternatives have come to market. The nanoVNA (nanoVNA.com) is an open source VNA device that is only a $200 in price. Of course, the nanoVNA is not as sophisticated at the most expensive VNAs, but it is good enough amateur radio enthusiast to test their rigs without breaking the bank.

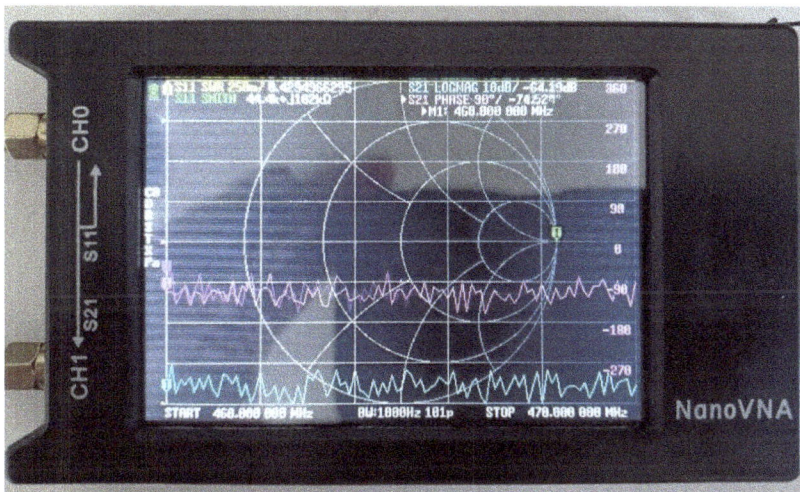

VNA can be 2 port (S11, S21) or 4 ports (S11, S21, S12, S22). The different ports are used to test the DUT forwards and backwards. The nanoVNA is only a 2 port VNA. There is also the NANOVNA LIBRECAL which is 4 port VNA. There have been different hardware versions of the nanoVNA. The hardware version for this text is 4.x, which has a 4-inch screen.

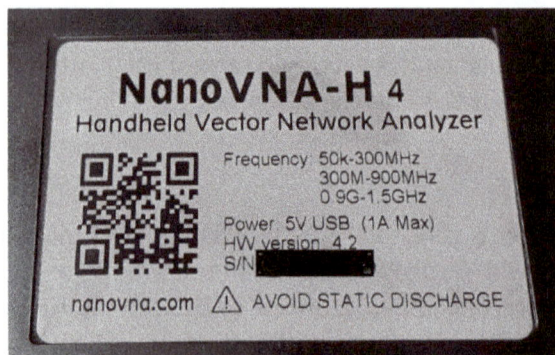

The nanoVNA comes with SMA connectors and calibration hardware. The calibration hardware are three SMA nuts that are open, short, and a 50Ω load.

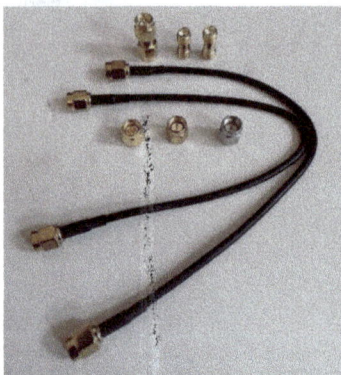

6.7.2 Software and Hardware Requirements

The nanoVNA is a nice self contains unit, but it is an open-source platform. The firmware for the nanoVNA is always getting upgraded. The nanoVNA.com site lists the software available, but for instructional purposes, these are the software tools needed:

1. STMicroelectronics DfuSe USB device firmware upgrade - https://www.st.com/en/development-tools/stsw-stm32080.html. The main chip in the nanoVNA is from STMicro. The tool updates firmware .DFU files into the chip. You will have to signup to gain access to the website to download the software.
2. NanoVNA-App – Reading the display on the nanoVNA is nice, but viewing the information on a large PC screen is even better. There are 3 software packages that interface with the nanoVNA over USB, but the NanoVNA-App provides the most complete solution for this explanation.
3. Firmware file to update the nanoVNA.

4. Calibration hardware (optional) – you may want to purchase additional calibration SMA nuts, cables, and connectors as these can get lost. Also, the extra load is a good thing for the isolation calibration step.

Like the programming tools that come with the radio, all these tools run on a Windows PC.

The DFu programmer can be used to program the firmware, but the NanoVNA-App has this capability built in. The only reason to install the DFu software is to get the STM32 Bootloader driver. Alternatively, the STM32 Bootloader driver can be downloaded from the TinySA-App site: http://athome.kaashoek.com/tinySA/Windows/Drivers/Win10/. The following text will continue to demonstrate installing the DFu programmer.

6.7.3   Firmware Upgrade

For a new out of the box nanoVNA, the first thing you will want to do is to upgrade to the latest firmware. There are two developers of firmware for the nanoVNA: DiSlord and Hugen. For this exercise, we will go with the Hugen version. The firmware page for Hugen79 lists several versions of the firmware. The hardware has changed because of certain chips going end of life or for compatibility reasons.

Looking at the back of the nanoVNA, if the model label has a MS or ZK at the end, then you need to download the .DFU file for the MS or ZK version respectively. If there is no code at the end, then this is the SI version. For the label shown earlier, the system is an H4 that requires the SI DFU firmware. For this example, the NanoVNA-H4-SI_20250526.dfu will be downloaded and installed.

1. Download and install the STMicroelectronics DfuSe USB device firmware upgrade. The only thing we need from the STM32 software is the USB device driver for the STM32 Bootloader. You can use the DFu programmer, but the NanoVNA-app has this feature built in. The STM32 software will install under the C:\Program Files (x86)\STMicroelectronics directory.
2. Download and extract the NanoVNA-App to a folder on your computer.
3. Download the appropriate DFU file for your nanoVNA and place the .DFU file in the same folder as the NanoVNA-App.
4. With the nanoVNA H4 powered off, connect the nanoVNA to the PC using the USB cable.
5. We want to put the nanoVNA H4 into firmware upgrade mode, press down on the jog wheel and turn on the nanoVNA. The screen will come up in black indicating that the nanoVNA is in firware upgrade mode.
6. Open Device Manager, and you will see that there is a STM32 Bootloader with a yellow bang.

7. We need to install the driver for the bootloader. Right click on the ST32 Bootlloader, and select Update driver from the context menu.
8. A driver update dialog appears, select Browse my computer for driver.
9. In the next dialog, open the path the C:\Program Files (x86)\STMicroelectronics folder, and click Next. The driver should install successfully.

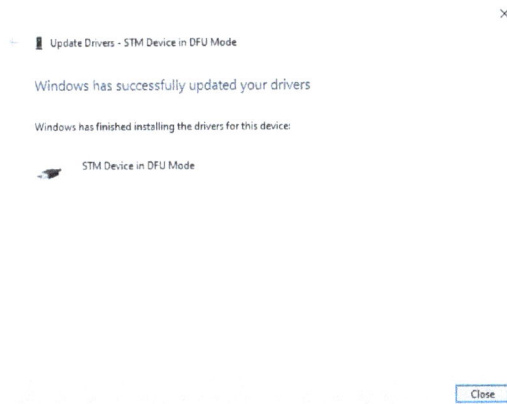

10. Click Close.
11. Run the NanoVNA-App.exe.
12. There is a button with two up arrows. This is the firmware upgrade button. Click on the button.

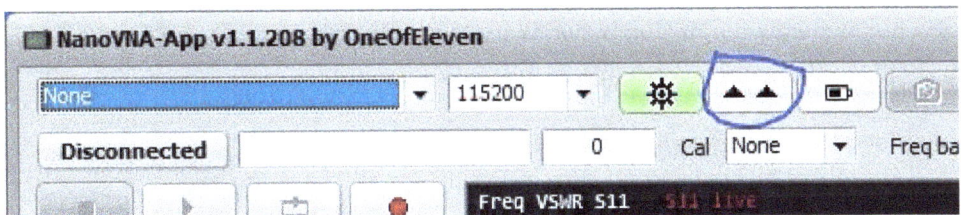

13. A new dialog appears, and automatically connects to the DFU device.

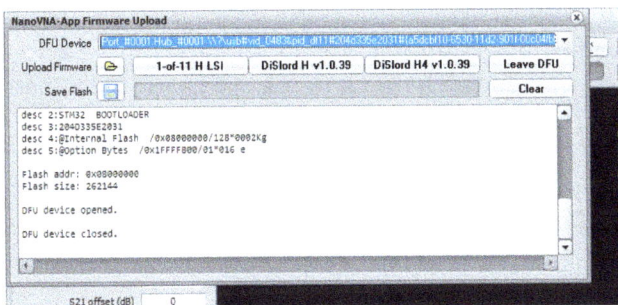

14. Click on the folder icon and open the .DFU file. The firmware will automatically be installed into the nanoVNA.

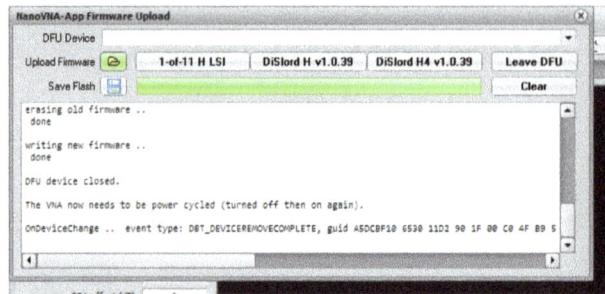

15. Power off the nanoVNA and close the firmware upload dialog.
16. Power on the nanoVNA. The unit is ready to be configured and calibrated for testing.

Note: Each time the firmware is upgraded, all previous settings will be lost. The nanoVNA will have to be calibrated for the frequency ranges under test.

6.7.4   Configure and Calibrate the nanoVNA for GMRS

When you first power on the nanoVNA, you will see four traces running. There are several steps to perform before we can test an antenna.

- Set up the traces to be displayed.
- Setup the Start and Stop Frequencies.
- Calibrate the nanoVNA for the start and stop frequencies.
- Test the antenna.

1. Power on the nanoVNA.
2. If this is the first power on you will see 4 traces that have four different colors. For testing the antennas, we want to see SWR and Smith Chart. Using the stylus, tap on the display to bring up the menu.
3. Click on Display -> Trace.
4. Double click on Trace 1 and Trace 3 to remove these traces. This should leave the Yellow and Green.
5. Click on Trace 0 so it is selected.
6. Click Back.
7. Click on Format S11 (REFL) -> SWR. The Yellow trace will now show S11 SWR at the top of the screen.

Note: Trace 2 should already be the Smith Chart. If not repeat steps 5-8 for Trace 2 and set to Smith.

8. Click Back.
9. Click Back to get to the main menu.
10. Next step is to set the start and stop frequencies of the sweep. For GMRS, 460 MHz to 470MHz covers all the GMRS frequencies. From the menu, click on Stimulus - > Start.
11. Enter the value 460.0 and click M for MHz.
12. Click on Stop.
13. Enter the value 470.0 and click M for MHz.
14. Click Back to get to the main menu.
15. Now, we need calibrate the nanoVNA for the selected frequences. Connect the long SMA cables to each of the ports on the nanoVNA.
16. Connect the SMA-F to SMA-F adapter to the end of the S11 cable.
17. Tap on the screen to bring up the menu.
18. Select Calibrate - > Reset
19. Select Calibrate at the top of the menu. The menu will show all the calibration steps in order: Open, Short, Load, Isolate, and Thru.

The three calibration nuts are used to calibrate the nanoVNA. In the picture from left to right is the Open, the Short, and the Load.

20. Connect the copper SMA nut that has no pin to the S11 lead.

21. Click Open button on the menu. The nanoVNA has a blue indicator progress bar at the top. It will check off the Open button when completed, and move to the Short button.
22. Remove the SMA nut with no pin.
23. Connect the copper SMA nut that has a pin to the S11 lead.
24. Click Short button on the menu.
25. Remove the SMA nut with the pin.
26. Connect the copper/silver SMA nut that is the load to the S11 lead.
27. Click on the Load button.

28. If you had a second SMA load nut, you would attach it to the S21 lead, but it you don't, with the load on the S11 lead, click Isoln button.
29. Remove the SMA load nut.
30. Connect the two leads together to make a loop between S11 and S22.

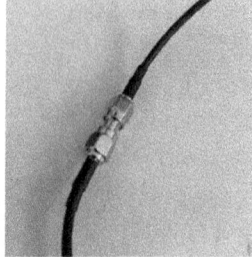

31. Click the Thru button.
32. When finished, Click Done.
33. You will be asked to save the calibration to a memory slot for future recall. Click on the first slot.
34. Tap on the display to close the menu.

The nanoVNA has been calibrated for the GMRS frequency ranges.

6.7.5   GMRS Antenna Test and Results

With the nano VNA calibrated, we can now test the different antennas. The S11 ports sends out the signal and then checks for any reflection (SWR). there was a DUT connected between S11 and S21, we would want to see S21 output on the graph. For antenna testing, S11 is all we need. If

1. Connect a walkie talkie antenna to the S11 lead.
2. Hold the lead at the base of the antenna so the antenna is vertical. If you leave the antenna horizontal the SWR will be very large, and not how the antenna would be used normall.
3. The nanoVNA will show you the sweep of SWR for the antenna from 460 MHz to 470 MHz.

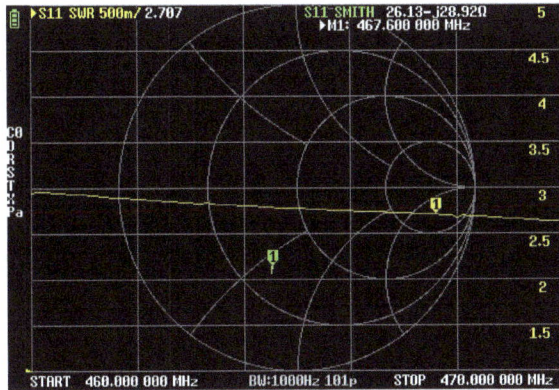

4. Use the jog wheel to move the marker up and down the frequencies. Stop at 462.6 and 467.6 to get the SWR value. You can change the scaling to adjust the SWR values per division. Notice that the slope / curve of the line matches the change in SWR per the results from the SWR meter.
5. Seeing the output on the nanoVNA is great, but let's see what this looks like on the PC. Connect the nanoVNA to the PC, and start the nanoVNA-App.
6. In the top left, select the COM port for the nanoVNA. The software should show Connected.

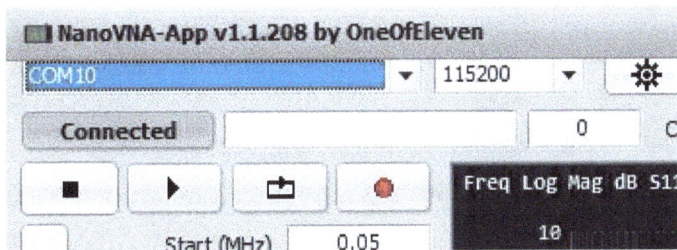

7. Set two markers at 462.6 and 467.6. In the bottom left cornet, left click on the Frequency maker box and select "Add marker" from the context menu.

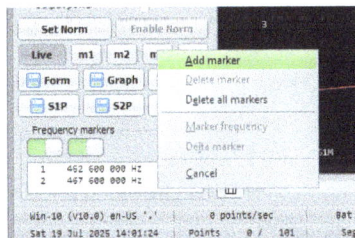

8. A New marker dialog appears, enter 462.6 and click OK.
9. Repeat steps 7 and 8 to set the 467.6 marker.

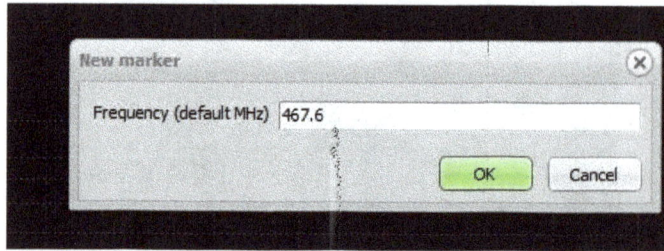

10. On the black graph, click on the LOGMAG title to bring up the context menu to change what the graph display.

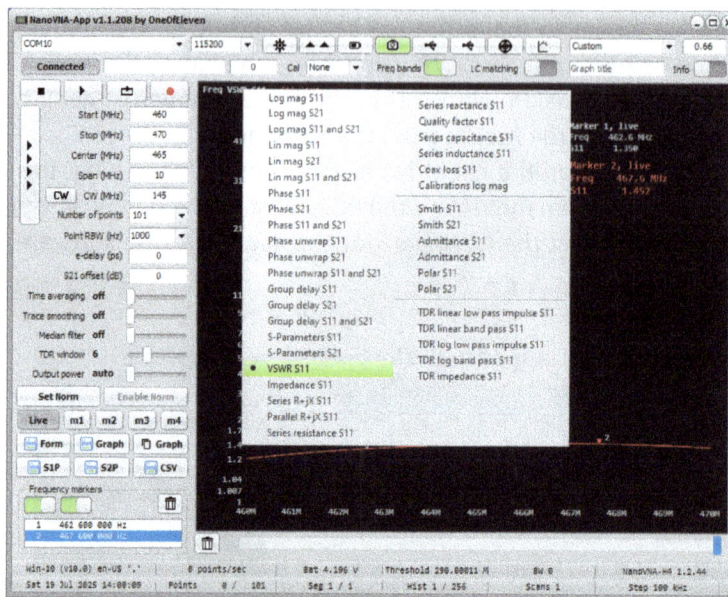

11. Select VSWR S11
12. On the left side set the Start to be 460 MHz and the Stop to be 470 MHz.
13. In the Cal drop down, select VNA to use the nanoVNA calibration settings.

Warning: Failure to set the Cal setting will have different results between the nanoVNA display and the nanoVNA-App.

14. Click on the Play button to perform a single sweep in the software. The SWR graph appears with the two markers and the SWE values.

Here were the test results that were performed with the available antennas:

114

Antenna DUT	462.6 MHz	467.6 MHz
KG-805G stock antenna	2.916	2.860
RT79P stock antenna	1.61	1.65
C2 stock antenna	2.92	2.07
RA86 stock antenna	1.62	1.66
Retevis M200	1.63	1.95
RT97L MA02 FRP	1.57	1.46
Retevis RHD-771 SMA-F	2.31	2.58
Retevis RHD-771 SMA-M	3.98	4.42
NKTECH NK-890G	5.78	7.6

Like the SWR meter we see that none of the results are a perfect 1:1 SWR. A SWR ≤ 1.50 is good. A little reliance on the internal antenna tuner is needed. The last two antennas in the list are a bit suspect.

6.7.6 Smith Chart and Antenna Tuning

The Smith chart is an important tool to help understand the impedance of the antenna. The multiple concentric circles can look intimidating at first, but once you learn how it works, it will make sense.

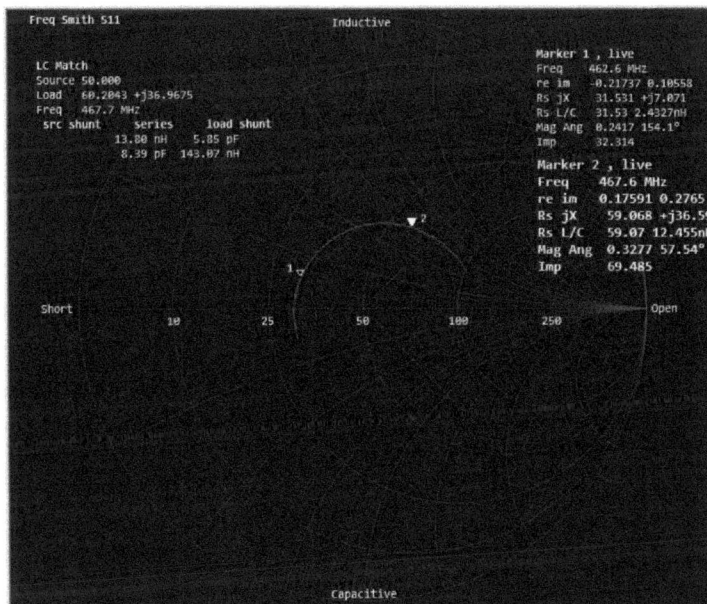

The line down the middle is resistance with no capacitance or inductance. On the left the line starts with a short 0Ω and the far left is ∞Ω. Right in the center is the 50Ω bullseye.

The top half circle is inductance and the lower half circle is capacitance. The picture above with the results from the Retevis M200 antenna over the frequency range. If you click on the LC match switch, capacitor and inductance suggestions appear on the chart. HAM radios have a tool to balance the load by using an antenna tuner to add capacitance or inductance. Antenna tuner could work for GMRS mobile stations. For mobile GMRS radios in vehicles, there are antennas that can be tuned to get the best impedance matching.

6.8   Harmonics

So far, we have covered the antenna side of testing. Now, were turn to an important test of the radio itself. More specifically, the output frequency and any harmonics. A harmonic is a positive integer multiple of the fundamental frequency. Musical instruments generate several harmonics when just play one note. For radio transmission, harmonics are generated by non-linear circuits. For example, let's say our fundamental frequency is 100 MHz. The fundamental frequency is the first harmonic.

> 2nd harmonic is 200 MHz
> 3rd harmonic is 300 MHz
> 4th harmonic is 400 MHz
> ...etc

For a GMRS radio transmitting on 462.5625, the harmonics are as follows:

> 2nd harmonic is 925.125 MHz
> 3rd harmonic is 1,387.6875 MHz
> 4th harmonic is 1,850.25 MHz
> ..etc.

Since the radio spectrum is used by other services, the radio cannot transmit the higher harmonics beyond the fundamental. FCC Part 95 subsection 95.1779 is very specific about removing unwanted emissions. The radio has to use a filter attenuate the high harmonics from being emitted into the air and creating interference. For GMRS radios, the output frequency spectrum can be tested to make sure the radio is rejecting the higher harmonics.

6.9   Spectrum Analyzer

A spectrum analyzer measures the magnitude of an input signal versus frequency within the full frequency range supported by the spectrum analyzer. Using a spectrum analyzer, we can see all the frequency on the output of a radio transmission.

6.9.1 TinySA Ultra

Like the VNA, a spectrum analyzer prices can range from $2,000 to $75,000+. Again not cheap for the average hobbyist. The good news is that the same developers of the nanonVNA have developed an inexpensive spectrum analyzer called the TinySA (TinySA.org). There are several editions and version of the TinySA. The original TinySA (TinySA Basic) came with a 2.8 in touch screen. The newer TinySA Ultra comes with a 4-inch touch screen. The TinySA Ultra has 3 model numbers ZS205, ZS406, and ZS407. The website covers the differences between the different models. Like the nanoVNA, the TinySA is not as sophisticated as the more expensive solutions, but it is good enough amateur radio.

WARNING! The TinySA is very sensitive to high power signals, which can damage the TinySA.

The TinySA is not just a signal analyzer, but it can also be used to signal generator. The TinySA Ultra has some additional features:

- Water fall feature to see the signal strength over time and what frequencies are be broad casted on.
- Ultra mode to scan frequency up to 12 GHz
- An external head phone jack, you can listen in on broad cast at different frequencies.
- SD card to store measurements.

6.9.2 Software and Hardware Requirements

The TinySA is a nice self contains unit, but it is an open-source platform. The firmware for the TinySA is always getting upgraded. The TinySA.org site lists the software available, but for instructional purposes, these are the software tools needed:

1. TinySA-App – Reading the display on the TinySA is nice, but viewing the information on a large PC screen is even better.
2. The STM32 Bootloader driver is needed to install the firmware. With the nanoVNA, we download and installed the DFu software to get to the driver. The driver is actually listed with the TinySA App: http://athome.kaashoek.com/tinySA/Windows/Drivers/Win10/. You will have to download the STtube.inf, Sttube.cat, and the .sys file in the x64 directory.
3. Firmware file to update the TinySA

4. Attenuator – the maximum signal strength input into the TinySA is 6dBm, which is 3.9mW. A 5W to 50W output signal from a radio is too much for the TinSA to handle. The attenuator used in the following sections is from BECEN 10W SMA 40dB, DC to 3GHz attenuator. Since the max power is 10W, only the 5W radios will be tested.

6.9.3   VNA Test of the Attenuator

The TinySA Ultra allows for 6 dBm max power input. There is going to be a little math. To convert watts to dBm, you can use the formula: dBm = 10 x $\log_{10}$(P/1mW), where P is the power in watts.

Walkie talkies output 5 watts or 37 dBm:

$$10 \times \log_{10}(5/1mW) = 37 \text{ dBm.}$$

6dBm is 4mW, which converts the other way as follows:

$$10^{\wedge}(6.0/10) \times 0.001$$

37 dBm is much greater than 6 dBm so an attenuator is needed to knock down the power. The attenuator used in for this exercise is rated for 40 dB, 10W, 0 to 3.6 GHz. The 40 dB attenuator will knock down the 5W output signal to -3 dBm or 0.5 mW. As a double check since we don't want to damage the TinySA Ultra, the nanoVNA can be used to test the attenuator. The following steps assume that you have walked through the nanoVNA steps in section 6.7.

1. Power on the nanoVNA
2. Change the traces shown on the screen. It you were measuring SWR and Smith using Trace 0 (Yellow) and Trace 3 (Green) respectively, hide these traces and show Trace 1 (Aqua).
3. Trace 1 should be set to S21 LOGMAG 10dB. If it is not, change the format of the trace to S21 LOGMAG.

4. Under Stimulus set the Start value to 50 kHz and the Stop value to 1.0 GHz. This is a good enough range that covers the GMRS frequencies.
5. Calibrate the nanoVNA for this new frequency range and configuration.
6. Save the settings to the second memory slot.
7. Connect the S11 port to the attenuator input port.
8. Connect the S21 port to the attenuator output port.

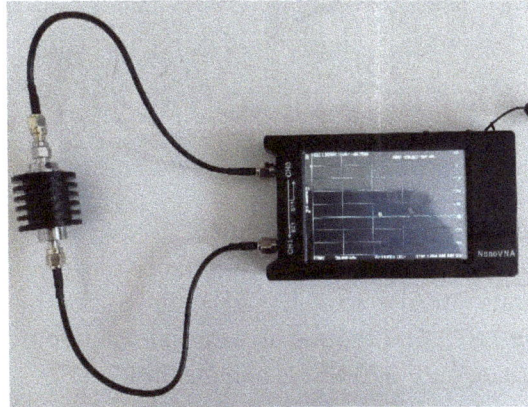

You should see the display with the trace line at -40dB from 50Khz to 1GHz, which means the attenuator is good. The readings distort near 1GHz, which is a nanoVNA artifact.

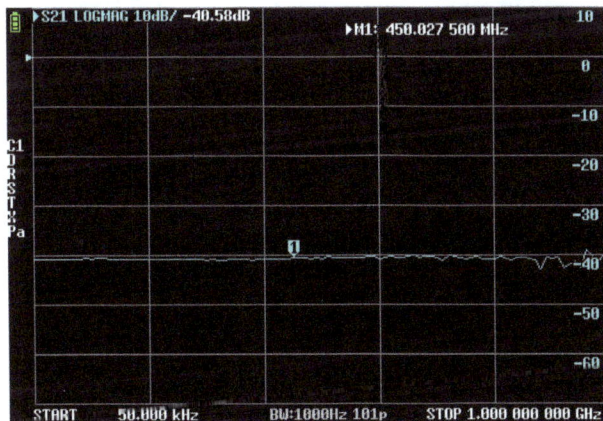

6.9.4   Self-Test and Calibrate the TinySA Ultra

Setting up and calibrating the TinySA is much simpler compared to the nanoVNA.

1. Power on the TinySA Ultra
2. Using one of the SMA cables, connect CAL port to RF port.

3. Tap on the screen to bring up the menu.
4. Select Config -> Self Test. The TinySA will run through a number of tests, and all the tests should be passing.

5. Once completed, select Config -> Level Cal.
6. A screen will appear to ask for which frequency range you want to calibrate to. Tab on "Calibrate 100kHz to 7.370 GHz" (6GHz for the other models). The Calibrate above 7.370 GHZ require an external signal source on the RF port. The frequency range below 7.370 GHz is fine for GMRS.

7. You will get another screen, just tap Calibrate to continue. The calibration process takes a few minutes.
8. The calibration should run to completion. Tap the screen to exit.

6.9.5 Firmware Upgrade

The TinySA.org website suggested to run through the self-test and calibration to make sure you don't have a damaged unit nor a knockoff version before doing a firmware upgrade. The Self-check and calibration will have to be performed again once the firmware has been updated. Here are the steps to upgrade the firmware.

1. Download the tinySA-app.exe file from the TinySA.org website. You will also notice a "Drivers" folder that contains the STM32 Bootloader driver.

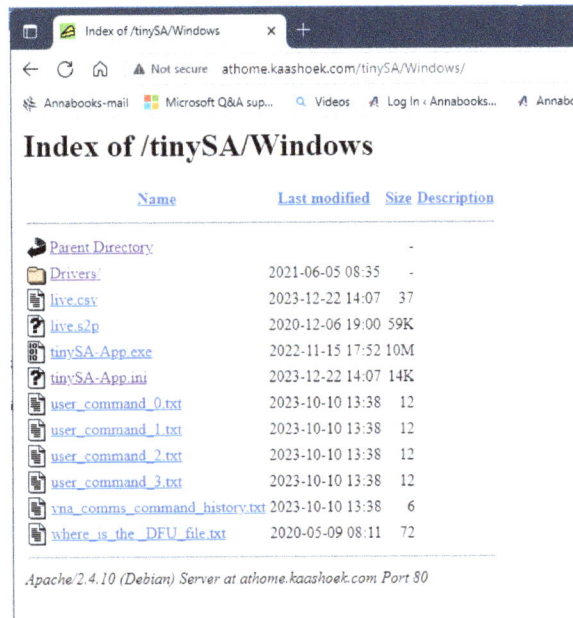

2. Place the tinySA-app.exe in a folder named tinySA on your computer.
3. Download the correct TinySA .DFU firmware file from the TinySA.org website. There is the TinySA and the TinySA Ultra. The tinySA4 (4 meaning 4 inch display) is for the TinySA Ultra.

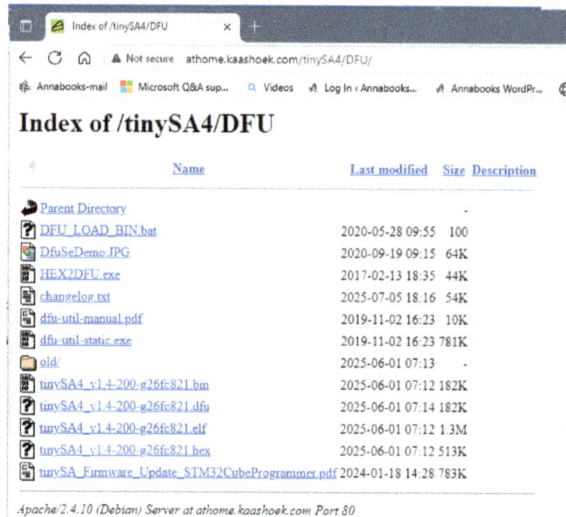

Index of /tinySA4/DFU

Name	Last modified	Size	Description
Parent Directory		-	
DFU_LOAD_BIN.bat	2020-05-28 09:55	100	
DfuSeDemo.JPG	2020-09-19 09:15	64K	
HEX2DFU.exe	2017-02-13 18:35	44K	
changelog.txt	2025-07-05 18:16	54K	
dfu-util-manual.pdf	2019-11-02 16:23	10K	
dfu-util-static.exe	2019-11-02 16:23	781K	
old/	2025-06-01 07:13	-	
tinySA4_v1.4-200-g26fc821.bin	2025-06-01 07:12	182K	
tinySA4_v1.4-200-g26fc821.dfu	2025-06-01 07:14	182K	
tinySA4_v1.4-200-g26fc821.elf	2025-06-01 07:12	1.3M	
tinySA4_v1.4-200-g26fc821.hex	2025-06-01 07:12	513K	
tinySA_Firmware_Update_STM32CubeProgrammer.pdf	2024-01-18 14:28	783K	

Apache/2.4.10 (Debian) Server at athome.kaashoek.com Port 80

4. Place the .DFU file in the tinySA folder along with the tinySA-app.exe.
5. With the TinySA Ultra powered off, connect the TinySA to the computer using the USB cable.
6. Press down on the jog wheel and turn on the TinySA Ultra. The screen should be black, which means it is in firmware upgrade mode.
7. Open Device Manager, and make sure the STM32 DFU Bootloader driver is installed. If the driver is not installed, you will have to download and install the driver manually.

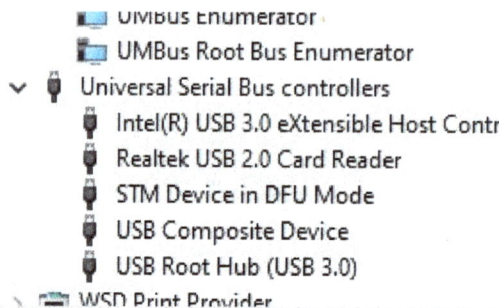

8. Open the TinySA-app.exe.
9. Click on the double up arrows to open the firmware upgrade utility. The utility should automatically connect to the TinySA Ultra.

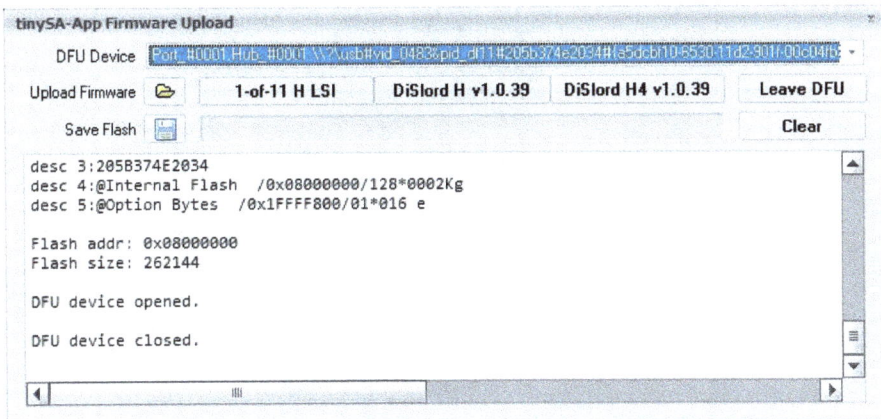

tinySA-App Firmware Upload

DFU Device	Port_#0001.Hub_#0001.\\?\usb#vid_0483&pid_df11#205b374e2034#{a5dcbf10-6530-11d2-901f-00c04fb: ▾			
Upload Firmware 🖿	1-of-11 H LSI	DiSlord H v1.0.39	DiSlord H4 v1.0.39	Leave DFU
Save Flash 🖫				Clear

```
desc 3:205B374E2034
desc 4:@Internal Flash  /0x08000000/128*0002Kg
desc 5:@Option Bytes  /0x1FFFF800/01*016 e

Flash addr: 0x08000000
Flash size: 262144

DFU device opened.

DFU device closed.
```

10. Click on the Folder icon and open the TinySA Ultra update .DFU firmware file. The update will start immediately and run to completion.
11. Power off the TinySA Ultra.
12. Close the firmware updater.
13. Power on the TinySA Ultra.
14. Run through the self-test and calibration again.

6.9.6   Radio Tests with the TinySA Ultra

With set up and calibration complete, we can now test the radio.

1. Power on the TinySA Ultra if not already powered on.
2. Since the 2nd harmonic for GMRS is near 1GHz, Ultra mode will be enabled. Tap on the screen so the menu appears.
3. Tap on Config -> More-> Enable Utlra.
4. A message will appear on the screen to enter a code. The code can be found on the tinySA.org website for enabling Ultra Mode. Tap on the message.
5. Enter the code.
6. The frequency range will go from 0Hz to 3GHz.
7. Let's set the start frequency. Tap on the screen to bring up the menu.
8. On the main menu, tap on Frequency -> Start
9. Enter the start frequency for 450.0 M.
10. On the main menu, tap on Frequency -> Stop.
11. Enter the start frequency for 1.0 G.
12. Connect an SMA cable between the output of the attenuator and the RT port of the TinySA Ultra.
13. Connect an SMA cable between the input of the attenuator and the radio antenna port.

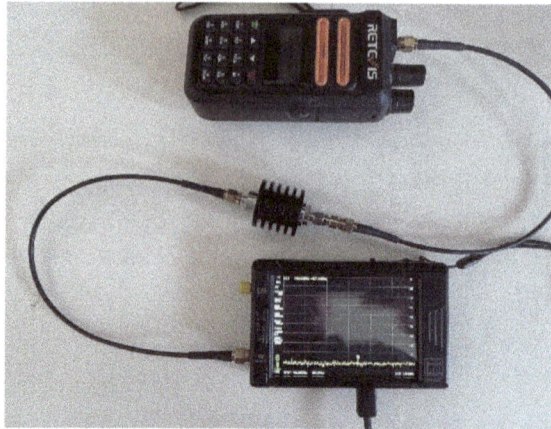

14. Turn on the radio and set the channel to 12, which is one of the low power channels.
15. Press and hold the PTT button on the radio until you see a peak appear on the TinySA Ultra. Takes to see a signal as the signal analyzer sweeps through the frequencies.

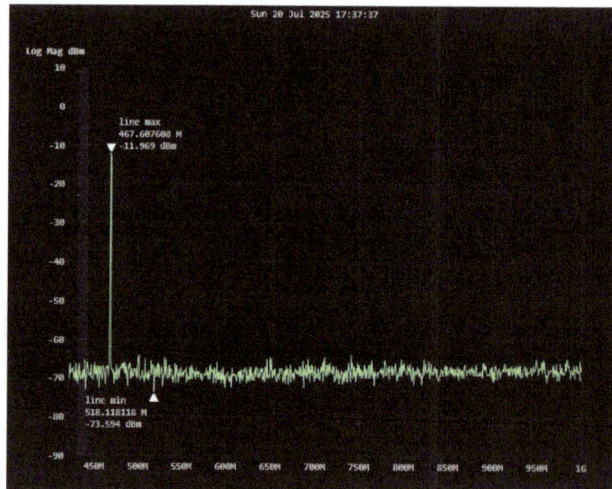

In the picture above, there is a peak at 467.6 MHz, which is near channel 12. The resolution one the sampling gets close enough. Most important is that there are no other peaks visible. If the radio had a bad filter, we see the 2nd harmonic appear at 935.325 MHz.

16. In the radio, change the channel to channel 15.
17. Set the channel to low power. You could leave the power high, but no need to emit that much power for this test.

18. Press and hold the PTT button on the radio until you see a peak appear on the TinySA Ultra.

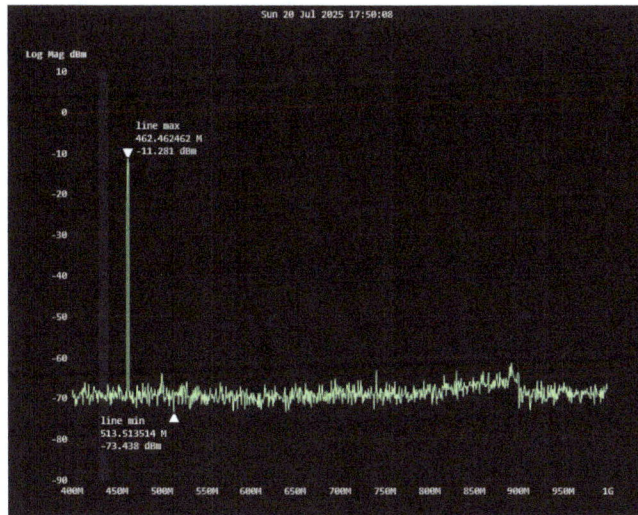

Again, there is only the fundamental frequency appearing in the analyzer. The radio is not emitting harmonics. If there were harmonics, the radio might need to be sent in for repair. The GMRS specification says that the max bandwidth is 20 kHz. We can test for the bandwidth, but it is better to use the application.

19. Connect the TinySA Ultra to the PC using a USB cable.
20. Run the TinySA-app.exe
21. In the top left, connect to the TinySA Utlra.
22. Keep the radio on channel 15, low power, and wide bandwidth (20 kHz).
23. In the TinySA application, set the start frequency to 462 MHz
24. Set the stop frequency to 463 Mhz.
25. Press and hold the PTT button and click on the play button in the TinySA-app. The sweep of the frequencies a will appear.

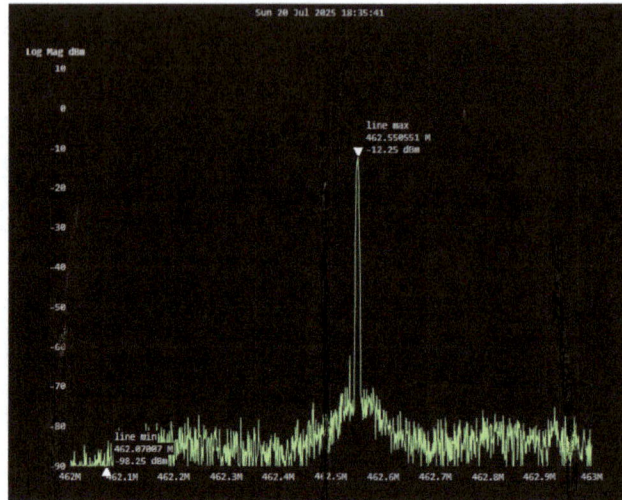

26. You can see the peak is very narrow. Use the mouse to check the lobes of the peak to verify the bandwidth is in spec.

6.10 Feedline Tests

Feedlines are used for mobile radios and repeaters. Damage to the feedlines can happen at any time. A break in the line can cause communication issues. A special tool called Time-Domain Reflectometer (TDR) cable tester is used to detect where the break is in the line. A TDR cable tester sends a signal down the line and measure the time it takes for the signal to be reflected back.

A continuity tester in a multimeter or a power meter can be used to check cables.

6.11 Where to Buy

The same stores that sell GMRS radios offer radio test equipment as well. Both the nanoVNA and TinySA sites warn of knock off duplicates that are not as good. Amazon has some of the test equipment. There is one place to get the nanoVNA and TinySA in the USA and that is R & L Electronics

Online Store	Website
R&L Electronics	www2.randl.com

If you have the money, there are spectrum analyzers with VNA capability like the Rigol RSA2015N or the SIGLENT SVA1015X.

6.12 Summary

GMRS provides a gate way to amateur radio. If you just buy a GMRS walkie talkie to communicate with other walkie talkies or repeaters, you don't have to dig any further. If you are a little curious, the RF concepts and tools discussed in this chapter are a little glimpse into the radio hobby world.

A Bibliography

Many references source were used for the book's development. The references are broken down by chapter.

Chapter 1 References

General Mobile Radio Service:
https://en.wikipedia.org/wiki/General_Mobile_Radio_Service

General Mobile Radio Service (GMRS) | Federal Communications Commission:
https://www.fcc.gov/wireless/bureau-divisions/mobility-division/general-mobile-radio-service-gmrs

Chapter 2 References

Video: What is RF? Basic Training and Fundamental Properties:
https://www.bing.com/videos/riverview/relatedvideo?&q=the+basics+of+radio&&mid=E0DFE05EB17ED60E979CE0DFE05EB17ED60E979C&&FORM=VRDGAR

How Radio Works: https://electronics.howstuffworks.com/radio.htm

Video: HackadayU: Introduction to Antenna Basics - Class 1:
https://www.bing.com/videos/riverview/relatedvideo?&q=the+basics+of+radio+antenna&&mid=A981DB111674E4773A76A981DB111674E4773A76&&FORM=VRDGAR

Make Your Own Low-Power AM Radio Transmitter:
https://www.sciencebuddies.org/science-fair-projects/project-ideas/Elec_p024/electricity-electronics/make-your-own-low-power-am-radio-transmitter

Radio spectrum: https://en.wikipedia.org/wiki/Radio_spectrum

Frequencies for Radio Controlled Vehicles in the US: https://www.liveabout.com/radio-frequencies-in-the-us-for-radio-controlled-vehicles-2862530

Multi-Use Radio Service: https://en.wikipedia.org/wiki/Multi-Use_Radio_Service

Chapter 3 Reference

Squelch: https://en.wikipedia.org/wiki/Squelch

Chapter 4 References

IP code: https://en.wikipedia.org/wiki/IP_code

Video: Every Radio Setting Explained & How it Works: https://www.bing.com/videos/riverview/relatedvideo?q=what+does+teh+R-tone+do+in+a+GMRS+radio&mid=0C24A641F08B42409F350C24A641F08B42409F35&FORM=VIRE

What are use-cases for DTMF & ANI sidetones in GMRS? (and related menus on KG-935G): https://forums.mygmrs.com/topic/3299-what-are-use-cases-for-dtmf-ani-sidetones-in-gmrs-and-related-menus-on-kg-935g/

Chapter 5 References

DB9 for RT97L Repeater: https://twowayradiocommunity.com/rt97l-repeater-faqs-solutions-for-your-common-questions/

Chapter 6 References

Video: You Have The WRONG Antenna - GMRS Radio Antenna Basics & 1st Look Midland Bull Bar Antenna: https://www.youtube.com/watch?v=hs1TVoAQ9K0

What is a Vector Network Analyzer and How Does it Work?: https://www.tek.com/en/documents/primer/what-vector-network-analyzer-and-how-does-it-work

GMRS Antennas and Understanding dB: https://midlandusa.com/blogs/blog/gmrs-antennas-and-understanding-db?gad_source=2&gad_campaignid=22440337608&gclid=EAIaIQobChMIoJy89MG9jgMV4yCtBh363SCyEAEYASAAEgJ4YPD_BwE

Video: Antennas Part I: Exploring the Fundamentals of Antennas: https://www.youtube.com/watch?v=ks8RQ9CD72Q

Video: The scariest thing you learn in Electrical Engineering: https://www.youtube.com/watch?v=pXWbdxOAuDs

Coaxial cable: https://en.wikipedia.org/wiki/Coaxial_cable

Understanding Sensitivity: https://www.monitoringtimes.com/html/sensitivity.html

What is Gain?: https://www.amateur-radio-wiki.net/gain/#:~:text=The%20area%20between%20the%20isotropic%20radiator%20circle%20and,SD%20%28W2LX%29%2C%201986%2C%20the%20Radio%20Amateur%20Antenna%20Handbook

Antenna db Gain Chart: https://g7lrr.com/antenna-db-gain-chart/

A Tutorial on the Decibel: https://www.arrl.org/files/file/A%20Tutorial%20on%20the%20Decibel%20-%20Version%202_1%20-%20Formatted.pdf

Gain (antenna): https://en.wikipedia.org/wiki/Gain_(antenna)

B Index

www.ingramcontent.com/pod-product-compliance
Lightning Source LLC
Chambersburg PA
CBHW081818200326
41597CB00023B/4295